Paul K. Feyerabend: Knowledge, Science and Relativism
Philosophical Papers Volume 3

This third volume of Paul Feyerabend's philosophical papers, which gathers together work originally published between 1960 and 1980, offers a range of his characteristically exciting treatments of classic questions in the philosophy of science. It includes his previously untranslated paper 'The Problem of Theoretical Entities', and the important lecture 'Knowledge without Foundations', in which he extends the perspective on early philosophy and science put forward by Karl Popper. Other themes discussed include theoretical pluralism, the nature of scientific method, the relationship between theory and observation, the distinction between science and myth, and the opposition between 'rationalism' and relativism. Several papers from the 1970s detail his increasing preoccupation with the social status of science and with the decline (as he perceived it) in quality within the philosophy of science itself. The volume is completed by a substantial introduction and a comprehensive list of Feyerabend's works.

John Preston is Lecturer in Philosophy at the University of Reading. He is the author of *Feyerabend: Philosophy, Science and Society* (Polity Press, 1997) and the editor of *Thought and Language* (Cambridge University Press, 1997).

Paul K. Feyerabend: Knowledge, Science and Relativism

Philosophical papers
Volume 3

Edited by JOHN PRESTON

CAMBRIDGE
UNIVERSITY PRESS

CAMBRIDGE UNIVERSITY PRESS
Cambridge, New York, Melbourne, Madrid, Cape Town, Singapore,
São Paulo, Delhi, Dubai, Tokyo, Mexico City

Cambridge University Press
The Edinburgh Building, Cambridge CB2 8RU, UK

Published in the United States of America by Cambridge University Press, New York

www.cambridge.org
Information on this title: www.cambridge.org/9780521641296

First published 1999

A catalogue record for this publication is available from the British Library

ISBN 978-0-521-64129-6 Hardback
ISBN 978-0-521-05727-1 Paperback

Contents

Acknowledgements   *page* vii

Introduction to volume 3
JOHN PRESTON   1

1   The problem of the existence of theoretical entities   16

2   Knowledge without foundations   50

3   How to be a good empiricist: a plea for tolerance in
    matters epistemological   78

4   Outline of a pluralistic theory of knowledge and action   104

5   Experts in a free society   112

6   Philosophy of science: a subject with a great past   127

7   On the limited validity of methodological rules   138

8   How to defend society against science   181

9   Let's make more movies   192

10 Rationalism, relativism and scientific method   200

11 Democracy, élitism, and scientific method   212

Appendix: The works of Paul Feyerabend
ERIC OBERHEIM   227

Name index   252
Subject index   255

Acknowledgements

'The Problem of the Existence of Theoretical Entities' originally appeared as 'Das Problem der Existenz theoretischer Entitäten', in E. Topitsch (ed.), *Probleme der Wissenschaftstheorie* (Vienna: Springer-Verlag, 1960), pp. 35–72. It is reprinted here with the kind permission of Springer-Verlag. The translation is by Daniel Sirtes and Eric Oberheim, of the Zentrale Einrichtung für Wissenschaftstheorie und Wissenschaftsethik [ZEWW], Oeltzenstr. 9, D30169, Hanover, Germany. They would like to thank M. Weber and N. Teubner for helpful suggestions, and the ZEWW which partially funded this work.

'Knowledge Without Foundations', © Oberlin College. Two lectures delivered at Oberlin College in 1961, under the auspices of the Nellie Heldt Lecture Fund.

'How to be a Good Empiricist: A Plea for Tolerance in Matters Epistemological' was originally published in B. Baumrin (ed.), *Philosophy of Science: The Delaware Seminar, volume 2* (New York: Interscience Press, 1963), pp. 3–39, and is reprinted here with the permission of the University of Delaware Press.

'Outline of a Pluralistic Theory of Knowledge and Action' originally appeared in S. Anderson (ed.), *Planning for Diversity and Choice* (Cambridge, MA: MIT Press, 1968), pp. 275–84. It is reproduced here by permission of the MIT Press.

'Experts in a Free Society' originally appeared in *The Critic*, volume 29, no. 2, November/December 1970 (Chicago), pp. 58–69. It has, unfortunately, proven impossible to contact the original publisher for permission to reprint it. It therefore appears here by kind permission of Grazia Borrini-Feyerabend.

'Philosophy of Science: A Subject with a Great Past' was first published in R. H. Steuwer (ed.), *Minnesota Studies in the Philosophy of Science, volume 5*

(Minneapolis: University of Minnesota Press, 1970), pp. 172–83. It appears here by kind permission of Grazia Borrini-Feyerabend.

'On the Limited Validity of Methodological Rules' was originally published as 'Von der beschränkten Gültigheit methodologischer Regeln' in R. Bubner, K. Cramer and R. Wiehl (eds.), *Dialog als Method* (*Neue Hefte für Philosophie*, Heft 2/3, 1972), pp. 124–71. It is reprinted here by kind permission of Vandenhoeck and Ruprecht publishers of Göttingen, Germany. The translation is by Eric Oberheim and Daniel Sirtes, of the Zentrale Einrichtung für Wissenschaftstheorie und Wissenschaftsethik [ZEWW], Oeltzenstr. 9, D30169, Hanover, Germany. They would like to thank M. Weber and N. Teubner for helpful suggestions, and the ZEWW which partially funded this work.

'How To Defend Society Against Science' first appeared in *Radical Philosophy*, volume 11, Summer 1975, pp. 3–8, and is reprinted here with permission.

'Let's Make More Movies' first appeared in C. J. Bontempo and S. J. Odell (eds.), *The Owl of Minerva: Philosophers on Philosophy* (New York: McGraw-Hill, 1975), pp. 201–10. It is reproduced here by permission of The McGraw-Hill Companies.

'Rationalism, Relativism and Scientific Method' originally appeared in *Philosophy in Context*, volume 6, ©1977, pp. 7–19, and is reprinted by permission of the Department of Philosophy, Cleveland State University, Cleveland, OH 44115, USA.

'Democracy, Élitism, and Scientific Method' is reprinted from *Inquiry*, volume 23, 1980, pp. 3–18, by permission of the Scandinavian University Press, Oslo, Norway.

An earlier version of Eric Oberheim's list of Feyerabend's works was published as 'Bibliographie Paul Feyerabends' in the *Journal for General Philosophy of Science*, volume 28, 1997, pp. 211–34. The English translation by Eric Oberheim appears here with the kind permission of Kluwer Academic Publishers.

Introduction to volume 3

The papers in this volume have been selected with several criteria in mind. Drawn from between 1960 and 1980, they all fall within the period spanned by the two previous volumes of Paul Feyerabend's *Philosophical Papers* (Cambridge: Cambridge University Press, 1981) and therefore do not overlap with the later phases of his work covered in *Farewell to Reason* (London: Verso, 1987) and *The Conquest of Abundance* (Chicago: University of Chicago Press, forthcoming). Some represent major landmarks in Feyerabend's thought, some of which have previously been published but have not before been available in English, others of which are now rendered less than easily accessible by their original place or mode of publication. Yet others are well-known and effective summaries of his views at different times. All of the papers are central to and representative of Feyerabend's work during the nineteen-sixties and seventies. All of them are linked together by overlapping themes, a few of which will be explored here.[1]

SCIENCE AND MYTH

One of Feyerabend's primary fields of interest was in the relationships between science and other forms of human thought and activity. In the lectures entitled 'Knowledge without Foundations' he argues that science, myths, and religious doctrines share many features, and that this refutes naïve empiricist accounts, according to which science started when people stopped speculating and started observing, or experimenting. General, explanatory theories, he argues, are nothing like the 'respectable' empirical generalizations prized by empiricists. Not only do they not merely summarize observational evidence, the best scientific theories even go *against* unanalysed experience. Only such theories allow us to analyse and criticize observation statements and experience, to strip away misleading appearances and make progress in science.

Empiricists might respond that science is at least *more* firmly rooted in

---

[1] References to publications of Feyerabend other than the papers reprinted in this volume, in the form '(19**)', are to the list of his works published here as Appendix. References to the two previous volumes of Feyerabend's *Philosophical Papers* appear in the form '*PP1*' and '*PP2*'. For helpful comments on this introduction, I am grateful to Eric Oberheim.

experience than alternatives such as myth. But Feyerabend insists that convinced believers can always provide empirical arguments for their theories, and that the best myths are as firmly based in experience as some highly appreciated scientific theories. A myth is not *merely* a subjective phenomenon:

> Far from being a figment of the imagination that is clearly opposed to what is known to be the real world, a myth is a system of thought supported by numerous very direct and forceful experiences, by experiences, moreover, which seem to be far more compelling than the sophisticated experimental results on which modern science bases its picture of the world. (p. 58)

So the idea that the distinction between myth and science lies in the amount of experience captured in the latter must be wrong. Partly because of 'the astonishing plasticity of the human mind' (p. 57), it is easy to find empirical support for apparently implausible ideas (such as the belief in witches and demons). But this means that neither the elimination of myths, nor the transition within science from one theory to another (e.g. the transition from Aristotle's theory of motion to Galileo's), can have been based on a return to the facts of observation (Aristotle's theory was itself firmly based upon observations). This became one of Feyerabend's central themes: theories always rig experiences in their favour. The support or 'positive evidence' they are afforded by observations is therefore suspect at best, and worthless at worst. The idea that theories are compared with one another for their ability to account for the results of observation and experiment is an empiricist myth which disguises the role of aesthetic, social, and 'irrational' factors in theory-choice. We might seem forced to admit that the only distinction between science and myth is the timing: science is the myth of today, myths are the scientific theories of ages past.

This conclusion is *not* yet drawn in 'Knowledge without Foundations'. There, as well as in 'Outline of a Pluralistic Theory of Knowledge and Action', Feyerabend followed Popper in supposing that there *are* features which distinguish science from myth, or at least *should* distinguish them, if science is to deserve a reputation for rationality and open-mindedness.

For Feyerabend, perhaps the primary danger attending the quest for knowledge is what has been called the '*myth predicament*'. This arises when theorists indulge in 'theoretical monism'. In 'The Problem of the Existence of Theoretical Entities' and elsewhere, Feyerabend argued in favour of the Kantian idea that the theories we subscribe to influence our language, our thought, and maybe even our perceptions. From this, he drew the implication that as long as we use only *one* theory in our dealings with reality, however empirically adequate that theory may be, we will be unable to imagine alternative accounts of reality. Just as purely transcendent metaphysical theories are unfalsifiable, so too what began as an all-embracing scientific theory offering certainty will, under these circumstances, have become an irrefutable dogma, a myth. The genesis of the predicament, as described in Section 6 of 'How to be a Good Empiricist', is as follows.

Certain theorists, in the grip of a favoured theory, begin to assume its adequacy. They become impatient with less successful alternatives to their theory, and refuse to consider them. Any further successes their theory accumulates reinforce these attitudes. By refusing to consider alternatives, they inadvertently shield themselves from situations which would constitute problems for, or *tests* of, their theory. They will have turned their backs on the only kind of evidence which matters: *negative* evidence, and it will seem to them as if there are no circumstances in which their theory could be refuted. Their determination to explain even recalcitrant facts in its terms will be re-doubled. And these efforts at explanation will apparently pay off. 'More than ever the theory will appear to possess tremendous empirical support' (p. 95).

However, this success will come at a high price. The theorists, by refusing to admit alternatives to their theory, will have diminished its empirical content:

> [T]his appearance of success *cannot in the least be regarded as a sign of truth and correspondence with nature.* Quite the contrary, the suspicion arises that the absence of major difficulties is a result of the decrease of empirical content brought about by the elimination of alternatives, and of facts that can be discovered with the help of these alternatives. (p. 95)

Truth, certainty, and hence knowledge, will apparently have been attained. But according to Feyerabend, what has been constructed will be a metaphysical system, a *myth*.[2]

A look at doctrines which we think of as paradigmatic myths will confirm this judgement. A myth, Feyerabend has already argued, is 'a system of thought, possibly false, perhaps very unsatisfactory from an intellectual point of view, which is imposed and preserved by indoctrination, fear, prejudice, deceit' (p. 64). It relates the truth as it sees it in a way which cannot possibly be mistaken; it is infallible. Myths are not incoherent collections of crazy ideas, but 'logical structures of great sophistication which ... remain intact in the face of almost every difficulty' (p. 64). The elements of a myth are related to each other in such a way that the result is its preservation, and even confirmation, under all possible circumstances. And this is done not just by ignoring difficulties, but by turning them to the theory's own advantage. If trouble arises in the attempt to apply it to reality, its human defenders are blamed. But even if we remove the

---

[2] Why, one might ask, in view of Feyerabend's repeated arguments that metaphysics has a positive function within science, should it *matter* if theories turn into 'metaphysical systems', or myths? I suspect that his reply at this time would have been that metaphysical systems, however necessary and valuable they may be, are still only *primitive* scientific products (ch. 3, p. 100), and that they only stagnate, but do not progress.

psychological forces which support the myth, this does not lead believers to realize that they have been the victims of a hoax. The power of a myth is not exhausted by purely psychological factors: it can give explanations, reply to criticism, and account for events which seem to refute it. It can do this, says Feyerabend, '*because it is absolutely true*' (p. 64) (or at least because it has 'the *semblance* of absolute truth' (p. 96, emphasis added)). The doctrine becomes 'known' *for certain*.

Nevertheless, neither the sense-certainty sought by empiricists nor the rational certainty sought by intellectualists is worth having. We can, if we wish, have 'certainty' in science, but it will be entirely man made and meaningless:

> At this point an 'empirical' theory... becomes almost indistinguishable from a myth. In order to realize this, we need only consider that on account of its all-pervasive character a myth such as the myth of witchcraft and of demonic possession will possess a high degree of confirmation on the basis of observation. (p. 95)

The use of closed systems of explanation like this is not restricted to 'primitive' societies. According to Feyerabend, there exist today very influential theories (parts of Marxism, psychoanalysis, and the quantum theory) which work according to similar principles. Being built in a way that enables them to take care of almost every difficulty that might arise, they make themselves safe from refutation. Thus, 'assumptions which are possibly true (but also possibly false) are interlocked inside a more comprehensive theory which, because of this particular form of collaboration and mutual support, will be absolutely true' (p. 66). But this 'absolute truth', like certainty, is worthless.

THEORETICAL MONISM VS. THEORETICAL PLURALISM

This antipathy toward the myth predicament was one of the main driving-forces behind Feyerabend's attempt to liberalise empiricism further than ever before. He thus devotes considerable time to showing how theoretical monism would retard the growth of knowledge. But what exactly *is* theoretical monism? Feyerabend's somewhat sweeping characterization raises several issues.

Sometimes, he identifies theoretical monism as a tendency, bolstered by psychological mechanisms, to restrict the number of theories in a given domain, for the purposes of inculcating a fixed belief in a particular theory. This he takes to be a sign that the theorists concerned 'have come to the end of their rope, that they can no longer think of any decisive objection ... and that they have therefore, for the time being, agreed to accept a single point of view to the exclusion of everything else' (p. 107). Feyerabend rails against this tendency by deploying arguments for theoretical pluralism, the 'method of classes of alternative theories' (p. 99), which implies that knowledge is best attained by working with a plurality of hypotheses. At other times, however, theoretical monism seems only to be the far more innocuous set of ideas that there is such a thing as truth, that some theories are closer to being true than others, and therefore that we can and ought to winnow out inadequate theories.

Feyerabend's rather selective way with some of the views he rejects means that there are many questions that need investigating here. First, he presents the theoretical monist as believing in the existence of a single 'correct picture of the world' or 'correct point of view' (p. 104). Does this mean a single true theory, a 'theory of everything'? One might well believe in truth, or truths, without committing oneself to the existence of such a thing, for although truths must be compatible with one another, they need not comprise parts of a *single* picture or theory. Or is Feyerabend, more radically, denying the existence of *any* truth, about *any* subject-matter? Then it would be hard to see *why* we should bother to be theoretical pluralists, what theoretical pluralism would be a means towards, and we surely would have lost 'touch with reality' (p. 110). One line of thought is that the objections to theoretical monism are ultimately based on humanitarian ethical values. Feyerabend tells us, after all, that 'the most decisive objection' to theoretical monism is that it is a method of deception and conformity:

> It enforces an unenlightened conformism, and speaks of truth; it leads to a deterioration of intellectual capabilities, of the power of imagination, and speaks of deep insight; it destroys the most precious gift of the young, their tremendous power of imagination, and speaks of education. (pp. 95–6)

Second, what does Feyerabend *mean* by appealing to 'knowledge' in his argument for theoretical pluralism? Pluralism, he assures us, affords us our best chance of securing knowledge. But are the goods what the customer ordered? In a famous passage, Feyerabend contrasts the monist's conception of knowledge with the kind of knowledge which results from pluralism:

> Knowledge so conceived is not a process that converges towards an ideal view; it is an ever-increasing ocean of alternatives, each of them forcing the others into greater articulation, all of them contributing, via this process of competition, to the development of our mental faculties. ('Reply to Criticism', *PP1*, p. 107. See also this volume, p. 184)

Could this be a conception of *knowledge* (as opposed to belief)? Does it, for example, involve relinquishing the ideal of truth, or replacing the usual 'absolute' conception of truth with a relativist conception? (Feyerabend does suggest that there are *different* notions of truth. But could the concept of truth be, as he puts it there, 'a relatively recent product'?) p. 200. Or does it take on board the sceptical view that we must continue to believe that there is truth, even though we must also believe that it is never knowable? Are developing our mental faculties, raising our theories to a higher level of articulation, and attaining a higher level of consciousness (along with the retention of our childhood dreams, and the other pleasing phenomena to which Feyerabend refers) supposed to be aims in themselves? Could we attain these things *without* yet affording us a better grasp on the truth? More generally, when Feyerabend talks of 'the improvement of our knowledge' (p. 106), and 'the progress of science' (p. 189) does he have in mind progress with respect to truth (or probability, reliability, etc.), or does he refer simply to the refinement of our affective sensibilities? There is a temptation to think that he might contest the opposition implied here, by insisting that we simply cannot separate cognitive virtues from affective ones. But this does not fit well with those passages where, at his most voluntaristic, he does explicitly oppose truth to other ideals which guide human life, and insists that we have to *choose* between them:

> My criticism of modern science is that it inhibits freedom of thought. If the reason is that it has found the truth and now follows it then I would say that there are better things than first finding, and then following such a monster. (p. 183)

To prevent or remedy the myth predicament Feyerabend first proposed his own 'positive methodology for the empirical sciences' (p. 80), suggesting that we ought to use a plurality of mutually inconsistent theories, playing them off against one another in an attempt to reveal their faults and limitations. This, he suggested, is the 'theoretical pluralism' the pre-Socratic thinkers (and subsequent successful scientists) used: they recognized the *human* origins of explanatory systems (and societies), and thus treated their own theories as eminently fallible *guesses*, guesses which would be improved as a result of a critical comparison with other guesses. One thing Feyerabend *always* opposed was 'demythologization', the attempt to weaken the literal content of theories by taking their claims as merely metaphorical (p. 143, note 15). The fruitful pluralism which generates progress, or just genuine plurality, simply doesn't work if we 'reinterpret' away the factual content of what we believe.

Sometimes Feyerabend mentions the idea, fostered within 'evolutionary epistemology', that the development of knowledge bears a strong analogy to the evolutionary development of species. He points out that 'the development of animal species is the result of a process of proliferation that goes on even if the existing species happen to be well adapted to their surroundings' (p. 106). But the evolutionary model does not fit well with Feyerabend's 'principle of tenacity', which urges us to retain theories, no matter how comprehensively they have been refuted: 'natural selection', after all, does require that some organisms reproduce less successfully than others, and therefore that their lineage goes to the wall.

PHILOSOPHY OF SCIENCE AS MYTH

Feyerabend also recognized that science and myth are related in another way: science, in virtue of being a complex set of human activities, inevitably generates and sustains its own myths. He came to see the very idea that there is a distinction between science and myth as itself one of the myths of science. However long he took to get there, it is clear that he did ultimately conclude that science *is* just the myth of today: '*Science is just one of the many ideologies that propel society and it should be treated as such*' (p. 187). In other papers, he pursued variations on this theme: science is just 'one of the many pastimes humans have invented to entertain themselves' (1967m, p. 413) or, much later, one of several 'supermarkets', just like art or religion, from which we select what we want (1994p, p. 146). One of the ultimate results of this assimilation is that the term 'myth' loses all its pejorative connotations: in 'Let's Make More Movies' (p. 198) Feyerabend calls for a revival of mythical ways of presenting theories.

'Outline of a Pluralistic Theory of Knowledge and Action' is one of Feyerabend's earlier presentations of the idea that science and philosophy could *learn* from the contemporary arts, which have managed to resist the vociferous demands of 'reason', and in which conformity to a single point of view is 'no longer demanded' (p. 105).[3] This paper should be read together with the slightly earlier 'On the Improvement of the Sciences and the Arts, and the Possible Identity of the Two' (1967m), where Feyerabend argues that, for the sake of our culture, the gulf between the arts and the sciences should *never* be closed. What excites him about modern art is precisely its pluralism: its use of 'the method of multiple representation' (1967m, p. 411). Freedom of artistic creation, he believes, could go hand-in-hand with 'the improvement of our knowledge' (p. 106). The real difference between sciences and arts, for Feyerabend, is not so much in their actual operations, for although science is now dominated by experts, instead of opportunistic dilettantes, pluralism 'played and still plays an important role in science' (p. 217). Rather, it is the received *ideology* of science, which is wholeheartedly monistic, that lets it down.

Feyerabend added something distinctive and important to this thought that science exudes its own mythological 'prose': he insisted that the myth that science is something other than myth arises as a result of the *philosophical* activity which parasitically attaches itself to science, rather than

[3] Feyerabend's fondness for expressionist theories of art can usefully be contrasted with Popper's vigorous critique of them (in his autobiography, for example: *Unended Quest: An Intellectual Autobiography* (Glasgow: Collins, 1976), esp. section 13). And Popper's view that it is in science where progress is most marked stands opposed to Feyerabend's idea that modern art supplies a better and more humane paradigm of intellectual activity, which can be used to generate a 'humanitarian science' (ch. 4, p. 110).

directly from the loins of science itself. Throughout his work, he refers to methodological 'schizophrenia', the split between actual scientific method and its philosophical reflection, which masks from us the true nature of science. I suspect that Feyerabend came to see himself not as an 'enemy of science', but primarily as an opponent of its contemporary *image*, as well as of the kind of science-dominated philosophy prevalent in the Anglo-Saxon intellectual world. He always reserved his most scathing comments for his own 'bastard subject', the philosophy of science, realizing, more vividly than certain other thinkers influenced by Ludwig Wittgenstein's later work, that science and scientists can be affected by philosophical rhetoric, that what started as an adventure of the human spirit could degenerate into 'big science', driven entirely by experts. Asked what he thought should be the task of the philosopher of science in a society in which scientific standards are treated like myths, Feyerabend replied, 'Not to add myths of his own to the myths of the scientists'.[4]

THE ROLE OF SCIENCE IN SOCIETY

According to 'How to Defend Society Against Science', the potential for liberation which science once possessed derived entirely from the fact that it overthrew existing belief-systems, and not at all from any supposed hold on or approach to the truth it may have had. (Feyerabend sometimes treated the possibility of a change in belief as an absolute value.) Like Popper, Feyerabend cannot tolerate the existence of what Kuhn calls 'normal science', that is, periods during which fundamental metaphysical and theoretical assumptions are taken for granted, in order to concentrate on the job of 'fitting' the paradigm to nature. Feyerabend's estimation of contemporary science sometimes seems even more negative than Popper's. The science of today, he tells us, is accorded a special status both by intellectuals, and by society. Intellectuals exempt it from criticism mainly because they think, wrongly, that it embodies a superior *method* for acquiring knowledge. And society lets it off lightly mainly because it is perceived, wrongly, to have a virtual monopoly on the production of desirable *results*. But these criticisms have to be set alongside the fact that it was usually the image of science, and the willingness of scientists to play up to it, rather than science itself, which fundamentally irked Feyerabend.

An English translation of the 1973 paper 'On the Limited Validity of Methodological Rules' is included here not least because, in his correspondence with Imre Lakatos, Feyerabend recommended it as 'the last good

---

[4] From an interview conducted by Teresa Orduña in Berkeley, California, March 1981.

thing I wrote'.[5] It also serves as an excellent summary of the line of argument in *Against Method*. Having reached that book's most famous conclusion, that there is no such thing as the scientific method, Feyerabend devoted a considerable part of his time to working out its political implications. He is one of a number of thinkers who seek to argue that democracy should not be confined to institutionalized politics, but should be massively extended to permeate the operations of everyday life. His favourite positive idea, in this connection, is that in a free society the activities and proposals of scientists should be evaluated by lay people and controlled by 'democratic councils' (chs. 5, 8, 11).

This 'Democratic relativism', presented here at most length in 'Democracy, Elitism, and Scientific Method', makes lively contact with the hopes and perceptions of many people who nowadays identify themselves primarily as members of one or another social movement, or grouping (or collection of such). The perceived pluralism of contemporary societies is a powerful temptation to follow Democratic relativism in protecting (by assigning basic rights to) traditions, *rather than* individuals. But, on reflection, this still seems deeply problematic. Do people who, by choice or otherwise, fail to identify with any tradition, simply fall between the cracks of this collectivist morality, becoming non-persons, devoid of rights? What of reformers and other misfits who try to *change* their tradition(s), from within: are they to be judged misguided as a matter of course? Or do they become 'the community of those who have nothing in common'? Or are there circumstances in which individuals count *as* traditions? And from where comes the confidence that democratic councils '*will* choose a democratic relativism as the basis on which their exceptions are imposed' (p. 223, emphasis added)? It is a serious question, surely not to be silenced by accusations of emotional blackmail or slander (p. 222), how such councils would differ from the workers' committees, soviets, kangaroo courts and, ultimately, show trials with which the recent history of 'democratic' socialist societies was blighted. Those who worry, legitimately, about the coercion exercised by the state upon science, or *vice versa*, certainly ought to consider whether there might not be a less potentially harmful way of de-coupling the two.

Feyerabend, to his great credit, certainly came to recognize that some of these problems had not been adequately dealt with here. In later work, he

---

[5] In an undated letter from the Feyerabend/Lakatos correspondence, now part of the Lakatos Collection in the Archive Division of the British Library of Political and Economic Science, at the London School of Economics. The article surely represents the zenith of Feyerabend's attempts to empower footnotes and to allow them to wrest control from the body of the text itself!

retreated from relativism specifically because he felt that it wrongly portrayed customs and traditions as fixed and autonomous entities.[6]

HISTORY OF THE PHILOSOPHY OF SCIENCE

In one of the papers reprinted here, 'Let's Make More Movies', Feyerabend compares philosophical accounts of the history of science with the possibilities for presenting that same history in a theatrical context. Not only did he have a lifelong interest in the theatre, he was once offered the job of production-assistant to Bertolt Brecht, and later wrote an important article on Eugene Ionesco (Feyerabend 1967b, 1967c). Here, he is keen to insist that an argument is far more than an abstractly representable train of reasoning, for it involves the behaviour, strategies and appearances of the disputants and onlookers. A theatrical presentation will not allow us to use only 'rational' criteria for evaluating an argument, it will force us to attend to the argument's 'physiognomy', and 'to *judge reason* rather than using it as a basis for judging everything else' (p. 192). Among the advantages of the stage presentation are that it provides concrete examples which 'guide the application' of terms like 'reason', and give content to the corresponding concepts (p. 195). Brecht's presentation of Galileo brings out the latter's use of tricks, as well as his insight. It emphasizes that when scientific thought leaves the monastery, turning from being something purely contemplative into being part of everyday life (pp. 193), the process involved goes *beyond* liberation, to the stimulation of insatiable appetites.

Feyerabend concludes that philosophical problems can be dealt with in better ways than by 'verbal exchange, written discourse, and, *a fortiori*, scholarly research' (p. 195).[7] He thus regrets philosophy's decision to restrict itself to the word, arguing that the degeneration of philosophy is manifest in the history of the philosophy of science.

Feyerabend rarely had kind things to say about the philosophy of science, or its professional practitioners. Consistently highest in his estimation came the pre-Socratic thinkers, the heroic figures of the original 'scientific revolution', the philosopher-scientists of the late nineteenth and early twentieth century, such as Hermann von Helmholtz, Ludwig Boltzmann, Heinrich Hertz, Ernst Mach, and Pierre Duhem, and those who followed in their footsteps, such as Max Planck, Albert Einstein and Niels Bohr. What are the grounds of these particular basic value-judgements? The scientific revolution, Feyerabend assures us,

---

[6] See the essays collected in *The Conquest of Abundance*.

[7] Who can say whether his call for thinking film-directors had any effect? It might have been interesting, though, to know his opinion of 'Badlands' (1974), directed by Terrence Malick, a philosopher trained in the phenomenological tradition, and translator of Martin Heidegger's *The Essence of Reasons* (Evanston, IL: Northwestern University Press, 1969).

is the *heroic time* of the philosophy of science. It is not content just to *mirror* a science that develops independently of it; nor is it so distant as to deal with alternative *philosophies* only. It *builds* science, and defends it against resistance and explains its consequences. (p. 196; cf. pp. 121 and 128)

'Pure' philosophers, by comparison, almost always came off worse (see, though, the remarks on Mill and Hegel in chapter 11, some of which also appeared as *PP2*, chapter 4). In 'How to be a Good Empiricist', they are presented as the new enemies of tolerance and scientific progress. However, Feyerabend's evaluation of just who counted as the villains of the piece varied. In papers reprinted here, the Vienna Circle and Popper get away relatively unscathed (except in chapter 8). These thinkers did not represent the nadir of the process, according to Feyerabend. After all, both the Logical Positivists' 'explications' of the language of science and Popper's criterion of demarcation could be seen, at first, to involve a *criticism* of what was considered to be scientific activity, resulting in the expulsion of metaphysics. Feyerabend usually reserves his worst ire for those post-logical empiricist productions which aim to make sense of scientific theories in terms of the schemata of formal logic (chapter 6, section 8; plus chapters 9 and 11). Where they seek only to logically reconstruct science, they are accused of conformism. And where they go further, to excise parts of science which do not meet their logical criteria, they are accused of misunderstanding it. Either way, they are judged to retard the process of scientific advancement. But, ironically, even the 'historical' philosophers of science, influenced by Wittgenstein's later philosophy, and with whom Feyerabend's own name is often associated, do not escape his withering tongue. Kuhn and his followers, for example, are referred to in these terms:

> Kuhn's ideas are interesting but, alas, they are much too vague to give rise to anything but lots of hot air. If you don't believe me, look at the literature. Never before has the literature on the philosophy of science been invaded by so many creeps and incompetents. (p. 185)

Feyerabend's complaint was that these philosophers had abrogated their duty to critically evaluate science, in favour of its mere description, 'clarification' or 'elucidation'. In their hands, philosophy of science had been separated from the live body of science itself. But in this situation the real content of science can no longer profit from its philosophical hand-maiden: the result can only be that science is embalmed, and hence retarded. One might wonder why Imre Lakatos, whose own proposed 'methodology' Feyerabend considered to be empty, was always accorded more respect. If, as Feyerabend concludes, Lakatos' work issues in no genuine methodological prescriptions, how can it 'be used to transform the process of science itself' (p. 137)?

In the nineteen-seventies and eighties, a number of contemporary philosophers of science (Feyerabend, Giedymin, Worrall, Cartwright, van

Fraassen, Zahar) rediscovered and reworked the thought of their discipline's major turn-of-the-century figures: Ernst Mach, Pierre Duhem, Henri Poincaré, Émile Meyerson, etc. (More recently, the project has moved on to resuscitating the reputations of Logical Positivists.)[8] 'Philosophy of Science: A Subject with a Great Past' was the first clear indication that it was Mach whom Feyerabend most actively sought to rehabilitate. His papers and remarks on Mach are among the most exciting and fertile products of the recent rethinking.

Partly because the Logical Positivists themselves claimed Mach as their founding father, his reputation as a rabid positivist (fuelled also by the Popperian tribe, and by Lenin) has raged unabated. Mach was a major influence on the early, as well as the later, Feyerabend. He reckoned Mach's admirers within the Vienna Circle unwanted friends, but he reserved his main fire for those who portrayed Mach as 'a short-sighted and narrow-minded positivist'.[9] The contrast with Popper, for whom Mach's opposition to the atomic theory represented 'a typical example of *the obscurantism of instrumentalism*',[10] is most notable. This particular Feyerabend project is pursued at most length in the later paper 'Mach's Theory of Research and its Relation to Einstein' (1984d). Feyerabend's Mach is a true philosopher-scientist, willing to criticize and change science on the basis of philosophical considerations and in the name of scientific progress. He is the very model of a philosopher who 'participates in the process of science itself' (p. 198), granting it autonomy from the criteria of logic, while evaluating and reforming it by reference to general humanitarian standards. It is also notable that Mach provides what Feyerabend identifies as 'a philosophy of science that completely abandons the idea of a foundation of knowledge' (p. 129), replacing it by a knowledge without foundations:

> A special feature of Mach's philosophy is that science explores *all* aspects of knowledge, 'principles' as well as theories, 'foundations' as well as peripheral assumptions, local rules as well as the laws of logic; it is an autonomous enterprise, not guided by ideas imposed without control from its own on-going research process. (pp. 216–17)

Several of the papers in this volume provide the illuminating spectacle of a thinker trying to walk a line between, on the one hand, the totally

---

[8] See, for example, N. Cartwright et al., *Otto Neurath: Philosophy Between Science and Politics* (Cambridge: Cambridge University Press, 1996).

[9] *Killing Time: The Autobiography of Paul Feyerabend* (Chicago: University of Chicago Press, 1995), p. 30. Note, however, that in 'How to be a Good Empiricist' (p. 83), Feyerabend does acknowledge some restrictive aspects of Mach's positivism.

[10] 'Three Views Concerning Human Knowledge', in his *Conjectures and Refutations* (London: Routledge, 1963), p. 100 note.

uncritical acceptance of science, which Feyerabend associates with the 'historical' philosophers of science and, on the other hand, the attempt to measure science too hastily by reference to abstract logical criteria drawn exclusively from philosophy. A bland historicist relativism, after all, will offer little scope for what Feyerabend consistently deems essential: the possibility of *criticizing* and thereby reforming science. But philosophies with maximum critical potential, such as the Popperian one from which Feyerabend defected, are apparently free to write off the actual history of science as an irrelevant matter of contingent fact. How much does it *matter,* for example, if the methodological rules which philosophers lay down are rules which scientists have not actually followed? What role does the 'detailed study of primary sources in the history of science' (p. 137) which Feyerabend calls for have to play in the *evaluation* of science?

Mach's work, I believe, represented for Feyerabend the most plausible resolution of this tension: Mach's 'critical-historical' approach is what he sought to emulate.[11] For both, it is important that non-scientists can come to understand, evaluate, and even make suggestions for improving science. Both sought to keep open the possibility of criticizing science *as a whole.* To this end, both sought to demystify the history of science, rescuing it from official textbook accounts and recovering it as what Feyerabend later characterized as a collection of 'historical' traditions.[12] Both were keen to disregard distinctions between scientific specialisms, and to appeal across such academic boundaries, using some parts of science to criticize others. Both recognized that there are episodes in the history of science where theories triumph because they have suppressed objections, and that such objections, as well as the alternative research programmes which inform them, can profitably be revived. Both regarded the revival of such research programmes, including indigenous 'folk' theories, as an important step towards the 'democratization' of science. And this political move they advocated in order to counter what they perceived as the over-extension of the authority of science, the elimination of individuality from scientific activity, and the threat of scientific indoctrination, resulting in an increasing tendency for science to become a 'church'.[13] This work raises in clearest form questions about the status of science *vis-à-vis* philosophy and issues

---

[11] For an illuminating account of this approach, and of Mach's influence on Feyerabend, see Steve Fuller, 'Retrieving the Point of the Realism-Instrumentalism Debate: Mach vs. Planck on Science Education Policy', in D. Hull, M. Forbes and R. M. Burian (eds.), *PSA 1994, volume 1* (East Lansing, MI: Philosophy of Science Association, 1994), pp. 200–11.

[12] For the distinction between 'abstract' and 'historical' traditions, see chapter one of *PP2*.

[13] See, for example, Feyerabend (1982f) (1983a) and J. Blackmore, 'Ernst Mach Leaves "The Church of Physics" ', *British Journal for the Philosophy of Science*, vol. 40, 1989, pp. 519–540. (This latter article should also be consulted as a counterweight to Feyerabend's interpretation of Mach.)

concerning the approach(es) to science a philosopher can legitimately take which, although all philosophers of science must perforce take a stand on them, cannot be said to have received any clear theoretical resolution.

Feyerabend's own attempted resolution is itself noteworthy, and takes us back to his critique of theoretical monism. He sought standards or criteria by reference to which science could be fairly evaluated. To think that such standards could be drawn from science itself, or that scientists are the only people who are competent to assess science, is to fall into the 'élitism' which he ascribes to Polanyi, Kuhn, Holton and Lakatos (chapter 11). Indeed, Feyerabend differs from almost all other twentieth-century philosophers of science in thinking that no purely cognitive or intellectual standards for evaluating science can be found. But this does not mean that he gave up on the attempt to evaluate science. He still sought 'standards which have the advantage of being simple, commonsensical, and accepted by all' (p. 127). How much this departs from Popper's 'critical rationalism', whose epistemology also explicitly rests on values,[14] is a good question. 'Knowledge without Foundations' illustrates, among other things, the sense in which Popper and Feyerabend both considered values to be the foundation of epistemology, and thence of philosophy. But it is not just the values appealed to that differ: where Popper insists on applying criteria of logic, Feyerabend (and the Mach he portrays) find such standards only in *ethics*. Feyerabend clearly rejects the idea that we should direct our lives according to 'scientific' reason. In 'Rationalism, Relativism and Scientific Method', one of his most in-depth examinations of rationalism, having distinguished several versions of that doctrine, he argues that even so, rationality is just one tradition among others. There, he expressly rejects rationalism in favour of a deeply voluntaristic relativism. Elsewhere he insists that 'there are no unambiguous scientific standards' (p. 212). But is this a rejection of rationality itself? Is there still room for interpreting Feyerabend as rejecting 'Reason' and 'Rationality', but not reason and rationality, and as urging a conception of rationality widened well beyond the confines of scientific reason?

A NOTE ON THE TEXT

On the rare occasions when Feyerabend's original text contained an obvious linguistic slip, typographical error or omission, any words I have changed or added appear in corner brackets. Apart from these, the only way in which I have altered Feyerabend's text is to flesh out the references he offers, giving full and uniformly presented citations for the books and

---

[14] See, for example, G. Stokes, *Popper: Politics, Epistemology and Methodology* (Cambridge: Polity Press, 1998).

articles to which he refers. The separate list of references in the original version of chapter 3 has been amalgamated into that chapter's notes, which have now been numbered. These changes I have not flagged. Notes added by the editor or the translators are marked as such, and appear in square brackets.

I

The problem of the existence of theoretical entities

Translated by Daniel Sirtes and Eric M. Oberheim

It is said that tables and chairs are directly observable, but that atoms, electric fields, and photons are not. What is meant by this is something like the following. In the case of tables and chairs, we make the transition from the perception to the thing and its properties quickly and without further consideration. Here, naïve realism is a psychological reality. But in the case of atoms, electric fields, and the like, such a direct transition is not possible. While a single glance is enough to determine whether the table in my office is brown, complicated measuring devices, as well as an interpretation of the readings of those devices on the basis of physical theories, are required if one wants to determine whether there are indeed electric fields present, how strong they are, and what properties they possess. This situation suggests the following *first explanation* of the distinction between observational and theoretical concepts: a concept is an observational concept if the truth-value of a singular statement containing either only that concept, or that concept along with other observational concepts, can be determined quickly and solely on the basis of observation, or at least if it is imaginable that a decision of this kind will be possible someday (the back side of the moon was observable, in this sense, even *before* the publication of the first picture). A concept is a theoretical concept if in order to determine the truth-value of a singular statement which contains it, theories, in addition to observations, are also required. To be brief and imprecise, an observation statement is accepted (or rejected) by merely looking (or listening, etc.). A theoretical statement is accepted or rejected by looking *and* thinking (calculating).

The problem of the existence of theoretical entities can now be formulated in the following way. Are there things to which theoretical concepts correspond (e.g. are there also electric fields in addition to chairs and tables), or must theoretical concepts not be conceived of as concepts which refer to existing objects? Mind you, this problem has been formulated under the assumption that the theory which contains the concepts in question is *true*. Therefore, this is not a problem which can be solved by scientific investigation (observations and the proposal of additional theories). We are presuming, of course, that scientific investigations have already achieved the best imaginable result; namely, the truth of the theory in question.

This last remark immediately rules out as insufficient both of the following two attempts to solve the problem. First attempt: the existence of theoretical entities is decided by observations in conjunction with certain theories; in the case of electrodynamics, for example, the observations in conjunction with Maxwell's equations. It is very easy to see that this answer only leads us in a circle. The application of Maxwell's equations to a concrete case can only lead to the assertion that in this concrete case an electrodynamic field was found if the problem of theoretical entities has already been solved in a positive sense, i.e., if we are already allowed to interpret the fundamental symbols of the equations being used realistically. Second attempt: the existence of the theoretical entities of a certain theory is decided by the relationship of the theory to another, more general theory. As an example, let us take a simple theory which explains free fall through force fields near the earth's surface. If this theory does not have additional consequences other than statements about gravitational acceleration at the earth's surface, then it will seem very doubtful whether the existence of new things has been asserted here. What seems to be the case here is simply a duplication of descriptions about *one and the same thing*, namely, about free falling objects. The theory of gravitation completely changes the situation. This theory explains the free fall, shows that strictly speaking the laws of free fall are false, and in addition explains a lot of other things. One can, it seems, say that the first modest theory was a vague anticipation of gravitational theory, and therefore that the term 'force' must be interpreted realistically *in it too*. Yet this presupposes that the term 'force' is already interpreted realistically in Newton's theory, and this is exactly the question that we want to solve in the problem of theoretical entities.

It follows therefore that neither the discussion of specific theories, nor the discussion of measurements in conjunction with theories can lead to a solution of our problem. What can solve or at least clarify the problem is a discussion of the nature of theoretical knowledge or, to use a less Aristotelian manner of expression, what can solve the problem is a discussion of scientific methodology.

Such a discussion, and with it the problem of theoretical entities itself, only makes sense if the following two assumptions are made. The first is that the existence of observable objects is not a problem, and that the existence of theoretical entities only comes into question because theoretical entities cannot be observed. It is this first assumption that actually distinguishes the problem of the existence of theoretical entities from the problem of existence altogether, and fundamentally simplifies its solution. The second assumption is that there are theoretical entities, and that not everything is observable. We begin with the discussion of the second assumption and will return to the first assumption later in this article.

2 DERIVATION OF AN APPARENT PARADOX

Thus, we begin with the discussion of the second assumption. Is it indeed the case that there are theoretical entities in the sense of the first explanation? Or, to draw on concrete examples, is it the case that elementary particles, fields, and the like are not directly observable, and furthermore that they will never be accessible to observation? To take the example of the gravitational field, we do *feel* the burden of a suitcase in our hand and we notice the gravitation (the gradients of some $g\mu\nu$) very clearly when we climb a steep mountain. We must also pay attention to the fact that an electrician can detect the voltage of an outlet or a battery very quickly and 'through observation' (i.e., without thinking about any theory). He uses his voltmeter, or even better, he uses his moistened finger or his tongue (for small voltages). He observes directly and makes no inferences (for example, he does not infer that 'the pointer has this or that position, therefore the spring which holds it at the outset has this or that tension, thus ... etc.'). Therefore, in the sense in which we introduced the predicate 'observable' above (first explanation), for the electrician, 'voltage' is an observational concept. Furthermore, if we look at the very instructive cloud chamber photographs, for example the first photograph of a positron, or Leighton's photograph of the decay of a μ-meson, can there still be a doubt that we possess a direct method of observing elementary particles? Now it is entirely conceded that not *every* scientific statement is decidable by observations in the way mentioned. An example in which direct observations are not yet available is the core temperature of a star or the weight of a newly discovered asteroid. In such a case, one first observes and then calculates, until finally after quite a long time the desired result comes to light. But if one takes into consideration how many things there are whose properties had to be inferred in a difficult way at first, which in the end were made accessible to direct observation (consider again the back side of the moon, or the distance between molecule centres in a crystal), then one will not attach too great importance to this fact. On the contrary, we can show that adherence to the principles of scientific methodology must lead, in the end, to the direct observability of all of the states of affairs which are asserted by the theory. Empirical method does demand, after all, that every assertion of physical theory be made accessible to verification by experience. It demands the construction of reliable and decisive testing procedures. Now let us assume that through adherence to this demand we have found a method of testing which leads to a very strict test, and therefore, in the case of a positive outcome, to a very certain criterion for the existence of a state of affairs *S*. As soon as this method is generally accepted and has been standardized, it is only a question of time until no conscious distinction is drawn between the presence of the criterion and the presence of *S* itself. The presence of the criterion no longer comes into consideration on its own, but one immediately says without further ado that *S* itself occurred: *S has become directly observable.*

The argument developed in the last paragraph can be summarized in the following way: many of the entities labelled theoretical are actually observable, and those which are not yet observable can be made accessible to observation. Thus, if we disregard the historical coincidence that certain methods of observation are not yet in use, we must conclude that *all descriptive concepts of science (or more generally, all empirical concepts) are observational concepts.* This contradicts the second assumption formulated at the end of the preceding section, and therefore the problem of theoretical entities as it was developed in that section dissolves into nothing. Now the real problem is no longer the question of whether and why we should interpret theoretical concepts (as opposed to observational concepts) realistically. The problem is whether and why we should interpret a descriptive concept realistically at all.

But for the time being, we are still a long way from the solution to *this* problem, because as every expert about the situation knows, there are a great number of objections to the result we just derived. Thus, before we proceed, we must clear the air by discussing each of these objections in turn.

We have said that the feeling of weight which we sense when we lift a suitcase can be conceived of as an observation of the gravitational field at the location of this simple 'experiment'. A very simple and naïve objection to this assertion, but one which is raised by certain philosophers, is the following. Lifting a suitcase is not at all an observation of the gravitational field. If anything, it is an observation of the weight of the suitcase; although even in this case – and even this remark is sometimes considered as a serious objection – the word 'observation' or 'experiment' sounds a bit artificial. Let us disregard this sideswipe at the artificiality of the expression! After all, this just shows us that problems of observation are not dealt with systematically in everyday life. This is a criticism of ordinary language, but not of the terminology that we used. We have to say, then, that even in everyday life, lifting a suitcase does not serve only to test the suitcase's weight. For example, after a long illness, we can lift a suitcase of a known weight as a test of our own strength and *not* as a test of the weight of the suitcase (which in this case is presupposed as known). Or, we lift the suitcase of a person who does not have a friendly disposition toward us, and we test his patience or our own nerve. One can increase these examples *ad infinitum.* What they show is that the object being observed depends on the *problem* present, and that this object is not already given by the simple act of observation. The problem of the intensity of the gravitational field at a certain location on the earth's surface is not formulated in everyday language – what one doesn't know, one doesn't speak about. But as soon as this problem is formulated there is the possibility of recruiting an entirely everyday action, like lifting a suitcase whose content is known (i.e., of known mass), for solving it 'by observation'. Thus, we *can* conceive of lifting a suitcase as an observation of the intensity of the gravitational field at the location of this action, and we can conceive of the sudden feeling that the weight of the suitcase has decreased as an observation of a sudden decrease of this intensity. (A more realistic example is the direct observation of a supernova by observing the sudden increase in brightness of a point of light in the sky.) This takes care of the first objection.

We turn now to the second objection. In this second objection, it is admitted that the feeling of weight that we sense when we lift a suitcase may play a role when we want to draw an inference about the intensity of the gravitational field. But it is objected that there is no sense in which we have a *direct* observation here, because we *infer* from the presence of the feeling to the field with the help of a physiological-physical theory, while in the case of an observation of a table, no such conscious separation between the act of observation, on the one hand, and the postulated object, i.e., the table, on the other hand, occurs. We simply see a table, but not a situation which could be conceived of as a (positive or negative) test of the presence of a table on the basis of theoretical considerations. This objection compares two different stages of learning: the blind man who just acquired the faculty of seeing will initially have difficulties inferring the present state of affairs, such as the presence of a table in front of him, from his impressions. After dealing longer with tables and other macroscopic objects, the distance between the perception and the object will gradually be reduced, until finally no phenomenological distinction can be drawn between what is perceived and what is supposed to exist – we directly perceive a table. There is not the slightest reason to assume that in the case of a gravitational field such a development cannot occur, and so we can immediately see how the other examples we used follow suit: the observation of electrons in a Wilson cloud chamber, the direct observation of a supernova with the naked eye, etc. The last example is especially instructive. An astronomer can become very well acquainted with a certain star which is barely visible to the naked eye through interferometric observations (determination of diameter), spectral photographs, and the like. One evening, he looks up to the sky and sees a strong increase in brightness. 'A supernova!' (directly observed!), he shouts, as he rushes to the spectroscope and the interferometer. In the spectrum, He-lines prominently appear: helium is expelled by the atmosphere. These lines have a thin, bright core:

the outer surface of the star is surrounded by a layer of hot gases which is followed by a denser, colder layer. The shift on the interferometer shows a diameter 500 times greater, and so on. Who would want to say that this is not a direct observation for the experienced astronomer?

At this point, a third decisive objection arises. It is to be admitted that astronomers, electricians, and physicists no longer infer certain things, if by this is meant that at the moment of observation (of the spectrum, the cloud chamber photograph, the voltmeter) a long series of mental operations are *consciously* carried out. Yet, the third objection continues, this *psychological* fact is irrelevant to the problem of the existence of theoretical entities. This is because if one calls on the astronomer to *justify* his quite directly obtained assertion, then he must nevertheless produce a whole series of theoretical explanations. For example, he will give explanations about the construction of the apparatuses used, as well as theories which allow him to interpret the reactions of these apparatuses in the way that he actually does. The truth of the theories used can never be conclusively guaranteed, so the existence of the partially observed, partially inferred entities can never be conclusively assured. The objection concludes that the concept of an electric field is a theoretical concept because every assertion about electric field intensities requires such a *justification* by theories whose truth is not assured, and this is totally independent of whether the statement was arrived at quickly or laboriously.

This objection, which in my view is the most important, obviously means that the distinction between observational concepts and theoretical concepts is no longer understood in the sense of the first explanation. Instead, it is now understood in the following sense (*second explanation*): an observational concept is a concept which is constructed so that a singular statement which contains only this concept is not only arrived at immediately, without reflecting on it at all, but is also a statement which does not require further *justification* other than pointing out that a certain observation was made. Observation statements are certain, not hypothetical. But as soon as a statement contains a theoretical concept, for its justification, in addition to observations, one must draw on certain specifications about instruments, theories, and so on. Therefore, the statement is hypothetical.

After proposing this second explanation, the problem of the existence of theoretical entities can be formulated again, in the following way: are there things to which theoretical concepts correspond, or must these concepts not be conceived of as concepts which refer to existing things? Mark you, the formulation of the problem now depends on the fact that we can never be sure of the truth of a theory. Therefore, it is again a problem which cannot be solved by scientific investigation, but one which requires a methodological analysis for its solution. In addition, such an analysis, and with it the problem of theoretical entities in the second formulation, only makes sense if the following two assumptions are fulfilled. As in the case of the first formulation, the first assumption again states that the existence of observable objects is unproblematic, and that the existence of theoretical objects is only questionable because they are not observable. The second assumption states that there are theoretical entities, and that not everything is observable. We will now show that the second formulation of the problem of theoretical entities collapses in upon itself, because the first assumption is not fulfilled, i.e., because all concepts are theoretical concepts in the sense of the second explanation.

Thus, for example, it is easy to show that the concept 'table' must be a theoretical concept. First, because the perception of a table from which we begin depends, after all, on our having properly learned to use a very complicated instrument: our eyes. Admittedly, we have received this instruction very early on, but it is still instruction, as is proven by the case of the blind person who becomes able to see only very late in life. Second, the perception of a table is dependent on the nature of the medium in between, as well as on the laws of the propagation of light through this medium. Third, the physiological state of the observer at the moment of observation plays a large role, and so on and so forth. That these factors all play a role in the justification of the assertion 'Here is a table' becomes especially clear when they are intentionally made explicit, i.e., in judicial hearings. But this just means that the determination of the truth-value of a statement about a particular table must draw on theories in addition to perception. That is, it turns out that 'table' is a theoretical concept. Furthermore, because this argument can be repeated with reference to any object, we must conclude *that all empirical concepts are theoretical concepts* (in the sense of the second explanation).

With this, we arrive at the result that the problem of theoretical entities collapses in upon itself whether one uses the first or the second explanation. This is because in the first case there are no theoretical concepts, and therefore no corresponding problem. And in the second case, *every* concept is problematic, because every concept is theoretical. No matter how one cuts the cake, it seems impossible to give a reasonable sense to the problem of theoretical entities.

We must now introduce a thesis that we have left out up until now, and which will throw a completely new light on the matter. This thesis, which is accepted by numerous philosophers, claims that there are concepts that exactly fulfil the criterion of observability in the second sense, even if the word 'justification' is taken very strictly. The thesis plays a role in different philosophical theories, of which the theory of sense-data is currently the most important. At this point, I would like very expressly to emphasize that I believe that the problem of the existence of theoretical entities stands or falls with the correctness of the theory of sense-data. If this theory, and especially the thesis contained in it, is false, then a reasonable sense for the whole question cannot be found at all, at least in the second formulation. Therefore it is necessary to investigate the correctness of the thesis mentioned with great care. This is what we will do in the next section.

3 SENSE-DATA

The fundamental assumption of the theory of sense-data is the following: there are empirical statements whose truth it is impossible to doubt under certain circumstances, and which are therefore to be considered absolutely true under these circumstances. The objects to which these statements refer are the sense-data. The descriptive concepts which appear in the statements are (directly) observable in the sense of the second explanation. Examples are statements about pains and smells, in short, statements about sensations, as well as statements of the form 'I perceive that ...' The truth of the statement 'I am now in pain' cannot, within reason, possibly be doubted by me when I am in pain. After all, I have immediate access to my pain, and I cannot even say what a doubt in this case should mean: what could be true other than that I am now in pain?

I do not intend to repeat the very plausible considerations which are stated for the existence of sense-data in this context. These considerations are well known to every student of philosophy. But are they faultless? This should be investigated more closely.

My criticism of the assumption of indubitable, *and therefore* absolutely true empirical statements mainly consists of three parts. First, I will show that many of the examples used, like pains, sensations of smell, and the like, are not sense-data in the sense in which we introduced this concept above. That is, I will show that statements about pains and other sensations are very often subject to doubt. Second, I will show that even in those cases in which doubt seems simply impossible, an inference to absolute truth is not allowed. Here, my most important consideration will consist in the fact that in the case of sensations, the impossibility (or the apparent impossibility) of a doubt is to be traced back not to the existence of absolutely binding reasons for truth, but rather to the impossibility of imagining alternatives. Naturally, these alternatives can always be excluded by convention, and therefore the meaning of an assertion about sensations can be defined so that it only really refers to what is immediately present. Such a convention, and this will be illustrated in the third part of my criticism, leads us to leave ordinary language, and in general, any language in which we speak about sensations, and it leads us to introduce an artificial language. This artificial or 'ideal' language really contains statements which are observational in the sense of the second explanation. It will turn out that such an 'ideal' language cannot serve as a means of communication, and therefore cannot be suitable as an observation language for scientific theories. Strictly taken, our rejection of sense-data is not based on the refutation of the absoluteness thesis, but rather on the decision not to use sense-datum statements because of their uncomfortable properties. I now begin with my criticism.

First of all, it is not true that *every* assertion about sensations excludes doubt. Anyone who has undergone a sense-test after a partial paralysis will know how difficult it is to distinguish whether a particular sensation is due to a pointed object or a blunt object. Mind you, what is doubtful here is not just a conclusion about the object, but also the correct identification of the sensation itself. Sometimes it is also very difficult to decide whether or not the sensation was painful. One first agrees, then withdraws agreement, and finally agrees without much conviction that the sensation was indeed painful. Even whether something was felt at all is sometimes doubtful. The explanation for this phenomenon is very simple. After all, a sensation is not something absolute. It is always a matter of contrast in relation to the background of other sensations which in general do not enter consciousness, and whose analysis therefore requires special preparations (one has only to consider the difficulties connected with the isolation and correct description of the subjective visual grey [subjectiven Augengrau] which David Katz described so exceptionally).[1] A very weak sensation may stand out a little bit from the pool of sensations for a moment, but may sink back into it again a moment later. This makes it difficult to decide whether something happened or not. Of course this applies not only to tactile sensations, but also to sensations of smell, visual sensations, and other sensations. It would be very instructive here to give more examples, especially the example of subjective visual grey. It seems to me, however, that the things already said suffice for us to propose the following claim: it is not true that statements about sensations can evade doubt *without exception*. I take this result already as a strong objection to the theory of sense-data.

But let's try to be fair! The sensations that we have described are exceptions, in so far as they are very weak. It is certainly not possible to doubt that I have pain when the pain is *very strong*. Strong sensations of smell cannot be subjected to doubt either. It also turns out that this assumption cannot be maintained. Even in the case of very intense sensations, it can be doubtful whether we are concerned with pains, noises, or other things. Imagine, for instance, that we are at the airport, just in front of an aircraft engine. The noise would be very uncomfortable, and even painful. Yet there is a point at which it is not clear whether noise is perceived, or whether pain is already experienced. Something very similar

---

[1] [See D. Katz, *Der Aufbau der Farbwelt* (Leipzig: Johann Ambrosius Barth, 1911), partly translated as *The World of Colour* (London: Kegan Paul, 1935). (Ed.)]

applies to an example which Berkeley discussed with a totally different purpose in mind. Berkeley correctly noticed that the feeling of heat at a high temperature transforms into a feeling of pain. Here, there is also a point at which one cannot decide whether the sensation is of heat or of pain. A third example is the case in which the sensation of pain gives rise to a feeling of pleasure (for masochists this is often the case, if not the normal case), and where, under certain circumstances, it is then no longer possible to determine whether intense pain or intense pleasure is present. Thus, it turns out that even intense sensations can be subjected to doubt. Given the ideas which we developed above, this also should not be surprising: if sensations are a matter of contrast, then the question of their appearance depends on the intensity of the background, and therefore becomes problematic as soon as the background sensations themselves surpass a certain intensity.

At this point, those who accept the theory of sense-data usually have two objections at hand. The first suggests that in these cases there is nothing more than a problem of description. A second objection suggests that what applies to atypical sensations does not have to apply to typical sensations (normal, strong toothaches) at all. We must examine these two objections more closely.

According to the first objection, the doubt can be traced back to the fact that there is no appropriate description in the language used for the present phenomenon. One uses the next best description, and feels that something is amiss, and therefore is not prepared to give one's full consent. Let us take, for instance, the example in which we cannot decide whether the sensation has the quality of pointedness or of bluntness. What we have in this case, a philosopher drawing on the first objection would say, is a kind of combination of the sensation that it is pointed and the sensation that it is blunt. The correct and appropriate description is therefore 'pointed or blunt', or 'dazzling-blunt', or the like. As soon as the correct description is found the problem is solved, and we have constructed a statement whose truth can no longer be doubted, even for this seemingly problematic case.

In order to reject this objection, we must first get clear about the expression 'evade doubt'. This expression can have a *logical* sense and a *psychological* sense. In the first case, it is emphasized, completely abstractly and on the grounds of the logical nature of the statement, that its correctness is absolutely certain, and this is totally independent of whether or not someone feels certain about it. Put the other way around, an assertion can be doubtful in the logical sense without this possibility weakening the *belief* in its correctness. I am completely convinced that I am now sitting at my desk, even if someone draws my attention to the fact that statements about desks are to be taken strictly as hypothetical, and therefore are not absolutely true. But a statement is *doubtful* in the psychological sense if one is not convinced of its correctness; if one's opinion fluctuates; and if one is not really sure what to say. Our examples obviously show that some statements about sensations are doubtful in the second, psychological sense. Is it now true that there is nothing other than a problem of description here? That is, is it true that statements especially constructed for these uncomfortable cases will be less doubtful? I think that this question has to be answered with a clear 'No'. If all *known* statements, whose sense and logic we are thus acquainted with, are regarded with mistrust, is it then to be expected that a new statement, subjected to our consideration for the first time, is going to be accepted with greater enthusiasm? The confidence with which we use a given statement in a perceptual situation is, after all, a result of training. If our training lets us down with respect to statements which we have used for a long time with great success, then it will fail us even more with respect to statements that we hear for the first time (by the way, this is an important objection to the credibility of unusual phenomenological descriptions). Naturally, it is true that long instruction in the use of this new description will finally lead to greater psychological certainty in their use. But in order to achieve this, first, *new* descriptions are not needed. And second, it turns out that it is not the existence of sense-data which regulates behaviour in perceptual situations. On the contrary, it is the compulsion to behave in a certain way which guarantees the existence of sense-data. (We will return to this very important point later.) That takes care of one part of the first objection.

Yet this does not completely dismiss the objection, because it can be argued that this is a logical problem and not a psychological problem. So here we go! There is a distinction between logical and psychological reasons for doubt! This presupposes that a statement about sense-data can be logically doubtful and psychologically certain and, vice versa, psychologically doubtful and logically certain. If one admits the first possibility, then one admits that the truth of a statement about sense-data depends not only on the immediate impression which one has at the moment of observation, but also on other factors, and with that one gives up the theory of sense-data. Yet the second possibility is even more of a riddle. It is assumed that a statement which cannot possibly be logically doubted is, nevertheless, looked at with discomfort. How is this possible if, as assumed by the theory of sense-data, all of the reasons for the truth of the statement are open to insight at the moment of observation? We cannot, therefore, – and this is the result of our considerations – draw a distinction between logical and psychological doubt in the case of sense-data, and it also follows that our argument can be considered as a complete refutation of the first objection.

The second objection admits that some sensations may be subject to doubt, but only in exceptional cases. And this is not valid for those cases which are at stake in test-conditions for scientific statements. For such test-conditions, clear, unequivocal sensations about which no doubt is possible are chosen, and *these*, not the cases discussed above, are to be considered as sense-data. Nobody can deny that there are such clear cases, and therefore, nobody can deny that there are sense-data. And this is totally independent of what may happen under more questionable circumstances.

This argument really has a strong persuasive force. Everyone must admit that sometimes pain is indubitably felt; that sometimes red is indubitably seen; and that these cases are free from the difficulties we have stated above. And thus it seems that one is forced to concede that in such cases doubt must really seem absurd and obstinate. With this, is the existence of statements which are indubitable, absolutely true, and which therefore are observation statements in the sense of the second definition proven?

It is very difficult to oppose an argument like this, which seems to unify the logical power of conviction with intuitive plausibility (who indeed can talk me out my pains?) with an equally strong, and most importantly, equally plausible argument. For this reason, my procedure will not consist in a direct attack, but in an attempt to undermine the opposition, to weaken it so as to prepare for its downfall. My first step in this manoeuvre consists in questioning from where the great certainty which we associate with statements like 'I am now in pain', and with other alleged sense-datum statements, comes.

We can attack this problem in two ways. We can discuss the psychological roots of this certainty, and its logical roots. Let us first look at the first case. We argued above that in certain exceptional cases even the appearance of very strong pain can be doubtful. It now turns out that an observer who has been exposed to a sufficient number of such special cases will critically inspect even normal cases. That is, even in the normal case, he will not say 'I feel pain' with the same certainty as before the instruction mentioned. Something similar occurs when one repeatedly asks an individual, who is not too obstinate, the same question: 'Is it really true that you feel pain? Are you sure that you're not mistaken?' If this individual does not happen to be a philosopher, then a point will arise at which he (or she) will no longer know what he (or she) should say, even in the case of very strong pain. According to the theory of sense-data, which does not allow for a distinction between what is immediately experienced and what is really present, such a case of course cannot show that *normal* pain can be doubted. Since a sense-datum theorist will emphasize that this entire procedure has led to a *change* of the phenomena and to the *removal* of normal pain, the sense-datum theorist will claim that we *talked* the subject of the experiment out of the sense-datum, and replaced it with other phenomena. This tactical move by the sense-datum theorist is of extraordinary importance, since it admits that the existence of sense-data is not something fixed for logical reasons, but is a question which must be settled by psychological investigation. In addition, it is also admitted that the existence of sense-data in a particular individual depends on that individual's history. Thus, one can create sense-data through appropriate treatment, and one can make them disappear again. Let us investigate the consequences of this concession.

For this purpose, let us assume that an infant A grows up in surroundings in which the word 'pain' is used very irregularly. Sometimes it is uttered in the presence of pain, at other times in the presence of smells, and at yet other times on the occasion of the performance of a modern opera, and so on. It is obvious that in such surroundings, no stable connection will arise between the word 'pain' and a certain psychological phenomenon, because there is no precisely describable phenomenon to which this word would be applicable. It follows that for A, 'I feel pain' will not be observational in the sense of the second explanation, and consequently *pain* will not be a sense-datum for him. Of course, this does not mean that A will not have any pains. Quite to the contrary, A's sensations are not fundamentally influenced by the situation described, yet these pains will no longer lead to the certainty in the production of statements which is characteristic of the presence of sense-data. This certainty, or the feature of indubitability which we normally associate with 'I feel pain', is therefore nothing more than a matter of *training* and of *regular instruction*. With that, first, we can understand why the demonstration of atypical cases can also shatter the certainty of normal cases. Such a demonstration turns out to be an instruction which disposes of the initially introduced regularity. (After all, one can learn and unlearn at any age.) Second, this explanation refutes the assumption of the sense-datum theory that *sense-data are the basis of theoretical knowledge, and are thus epistemologically as well as temporally primary*; and in addition that they are the only objects which can be said to exist with certainty. This is because an instruction which leads to the creation of sense-data can only be executed if the existence of pain, smells, and so on is *objectively* certain. But there is a lot more to it. Anyone who has looked through a microscope will know that in the beginning, not only the perceived physical object (for instance a bacterium), but even perception itself, is a very unstable and doubtful affair. First, one has to learn to see correctly in the psychological sense of the word in which sight does not include identifying an object at issue, but only the presence of a clearly structured and directly describable perception. What is the nature of such a perception after one has been trained to use the instrument? It corresponds more or less exactly with the object whose existence is claimed by the biological theories concerned. (For example, preformation theorists left behind drawings in which little human beings, little animals, and the like, are enclosed in a sperm. It is not very surprising that such observations are made. A microscope image is, after all, a very complicated matter, not dissimilar to a picture puzzle, and everyone knows that one can see anything there.) But that means that circumstances which are observable in the sense of the second explanation are the result of a training in which the existence of certain theoretical entities is assumed. Briefly and paradoxically put, *psychologically speaking, sense-data are the result of our belief in the existence of certain theoretical entities*. To dispose of such a belief, therefore, would lead not only to the disposal of our theories, but also to the disposal of the sense-data themselves, *except if one makes the assumption that there are innate concepts*.

So far, we have been concerned with the psychological character of sense-data, namely with the feeling of subjective certainty we have in their presence. Although this feeling is a *necessary* condition for the existence of sense-data (and when we discussed the 'existence' of sense-data, we always meant, of course, only this necessary condition), it is still not a sufficient condition. Or more precisely, psychological indubitability (whose existence we admit in some cases) does not need to have absolute accuracy as a consequence. As long as other conventions have not been adopted, it is completely undecided whether a statement which is indubitable, in a very simple and psychological sense, is therefore already absolutely certain and free from error. Or, expressed differently, absence of doubt is entirely compatible with hypothetical character, in the absence of other explanations. Later in this section, we will discuss the convention which is required in order to make an absolutely correct statement out of a psychologically indubitable statement, and we will then also show that this convention leads to consequences which abolish science as a systematic and inter-subjective enterprise. For the moment, we just want to show that such a convention is necessary *in addition* to psychological indubitability.

For this purpose, let us assume that we have sat at a certain table for a long time, that we ate our meal at it, and that in contravention of all good manners, we loll about it. Perhaps we dropped a pencil and had to retrieve it from under the table, and thus had a good look at the underside of the table. Under these circumstances, is it still possible to doubt the existence of the table? 'Sure', one might object, 'as we still have not executed all possible tests. For example, we did not try to determine what happens when it is dark, or what happens when it is in the presence of good-looking girls. It is conceivable that when the light is turned-off, all tactile sensations disappear, that all the glasses and a bottle of wine fall to the floor, and also that Professor Maxwell, who despite his good manners was sitting on the table, suddenly finds himself in the middle of the glasses on the floor. Such an experience would clearly show that the statement "A table stands in front of us" was incorrect, as persistence in darkness is a test-condition for the existence of tables. Therefore, it is still possible, despite all evidence stated, to doubt that the statement "A table stands in front of us" is correct.

And this statement, therefore, is also not absolutely true with reference to the evidence stated.'

This objection has a catch (and it is Professor Maxwell who drew my attention to it). How do we know, we could ask, that persistence in darkness is a *test-condition* for the existence of tables, and not rather a *coincidental property* which is often there, but not always? Therefore, what forces us to take the absence of tactile sensations in darkness as evidence for there having never been a table there, and not rather as evidence for the table at which we sat being a table which existed when there was light, but which *disappeared* when it was dark? One might be tempted to argue that this requirement is derived from physical theories about the persistence of matter in darkness. Yet this argument does not hold water. Because we have decided to interpret the described incidents assuming that the table was originally there and disappeared when it was dark, we must obviously then consider the physical theory at issue as refuted. Matter is not always insensitive to changes in illumination.

Some philosophers find their way out of this difficulty in the following way. They consider the persistence of certain sensations over a not too short period of time as already a totally sufficient criterion for the truth of the statement 'Here is a table.' Their argument is that with the word 'table', we just do not *mean* anything other than an object which leads to visual and tactile sensations of a certain form and persistence. As this is all we mean by the word 'table', the existence of a table is then already totally certain if the enumerated experiences are present. And it is therefore also possible to make assertions about physical objects which are indubitable and absolutely certain with reference to a particular class of facts. It is this argument which I will use in order to show the *logical* source of the great certainty of sense-datum statements.

First, I will make a small correction to this argument. It was hinted that we *mean* an object which has certain effects and not others by the word 'table'. (One should notice, by the way, that this argument is nothing other than the linguistic parallel to the older argument which is grounded on the *nature* of the things investigated. It is to be admitted that the transition to the linguistic way of speaking has a psychological advantage: one is more prepared to change a way of speaking than to admit that the *nature* of a thing is different than presumed. However, a glance at the linguistic movements in contemporary philosophy makes it seem doubtful whether a theory of language-use is less dogmatic than a theory of essences.) Yet, we would hardly change our designation if we were to discover or learn that the things which until now we called 'tables' are also the source of gravitational forces. There is also not the slightest reason why we should not draw on the existence of gravitational forces in testing assertions about tables. This example is, of course, a little bit unrealistic because we will hardly ever come upon a situation in which the measurement of gravitational forces would be a more convenient method for proving the existence of tables than direct observation with our eyes and hands. But with massifs, planetoids, or the dark companions of double-stars, the situation is just the opposite. Here, measurement by gravitational forces may even be the only applicable method. As soon as we succeed in giving a causal-physiological explanation of the influence of matter on our sense organs, or at least as soon as we have the idea that our sense impressions can be traced back to the causal influence of observed objects on our sense organs via an intervening medium, there is no longer the slightest reason arbitrarily to restrict the definition of the word 'table' to the production of sense impressions. We would then know that these sense impressions are only a part of the effects that tables have on their surroundings, and that these effects are distinguished only by the fact that most people are acquainted with them alone. Besides, such a definition would run counter to the principle not to use *ad hoc* explanations for the explanation of the phenomenon at issue. This is because its consequence is that the question 'How is it that we feel a resistance and we have a table-like visual impression?' is answered with a statement ('because a table stands here') which is logically equivalent to the description of the state of affairs at issue.

But let us grant the philosophers the freedom to propose senseless definitions, and therefore let us allow them to define the word 'table' so that sensations of a certain kind already guarantee the existence of tables! Then we will still have to say that the indubitability of statements about tables is to be traced back to exactly this *definition* which declares that the testing procedure is finished at a certain stage, and which claims that from now on the existence of a 'table' is certain. This is a very important point as it shows that decisions and conventions play a large role in questions of indubitability. Later, we will see that the alleged indubitability and absolute correctness of statements about sense-data is not something which has the 'nature' of sense-data as its foundation, but rather that it concerns the result of decisions. Our discussion will then be concerned with the *practicality* of these decisions, and that is all that remains of the problem of the 'existence' of sense-data. (By the way, one should notice that we have already recognized that the definition introduced above for 'table' is very impractical. It leads to *ad hoc* explanations about the appearance of sensations of a certain kind, and this applies generally to any kind of definition.)

But let us return again to the case of the table. We have claimed that it is a (very impractical) *decision* which guarantees the certainty of the result after a series of operations. Yet, one can counter this with the fact that we feel certain of the truth of the statement 'A table stands in front of me' in everyday life, even without explicit definitions. The explanation which some philosophers give for this is that we unconsciously rely on a certain definition, because after all we use the word 'table' unconsciously in a well-defined sense. I consider this explanation totally unrealistic. It assumes that in everyday life we make long-term decisions about our reactions to all possible, imaginable, and unimaginable experiments. Every 'ordinary person' who experiences hallucinations, mass-hallucinations, systematic illusions, and the like for the first time, would admit that this assumption is an illusion. He does not composedly explain, on the basis of 'implicitly contained definitions in the use of language', that he has just observed an unusual table, but rather he simply does not know what he should say. *In retrospect*, he can of course decide to stick with the statement 'Here is a table' even if he has been exposed to mass-hallucination and understood what was going on. What is important is that the statement is certain (according to the evidence incorporated in the definition) *just through this decision*, and it was not already certain beforehand (except of course in a purely psychological sense). This is because the latter would presuppose that one already knew which decision would be the one finally accepted.

We will illustrate this argument with another example. The ordinary person, so it is said, uses the word 'table' in a certain way, and from this it follows that he makes certain 'definitional' assumptions about the nature of tables. He does not make this plain in the form of explicit definitions. Rather, these assumptions are contained implicitly in his language, and they guarantee certainty. Well, what is the 'nature' of a table? In everyday life, we say that a table is something that we and others can see and touch. Therefore, if we see and touch something table-like, and if others experience the same sensations, then there cannot be the slightest doubt that it is a table. What is the logical reason for this strong conviction? Does it show that there is a logically indubitable statement present? In order to investigate this question, let us assume that a photographer takes a picture of a table, and that on the plate there is nothing to see except an empty room and two or three people who act as if they were sitting in front of a table. Let us also assume that nobody has any idea about the nature of photography. Wouldn't we then reject the photographic plate as totally irrelevant? On what grounds is this attitude based? I believe that this attitude is simply the result of not knowing ('Ignorance is bliss'). And this is how it is in most cases in everyday life. We are so sure of ourselves that we believe that neither decisions, nor further investigations are needed, simply because we have no idea of the situations which could force us into such decisions. Thus, we see that in everyday life the problem of the logical certainty of statements is still completely open, because we still don't understand it at all. We are sure of ourselves. This is correct. But from this, nothing at all follows about the absolute truth, or even the *truth*, of the statement which we claim with such certainty.

After this excursion, let us turn to answering the question of how philosophers motivate the definition stated above, as they would hardly even pose it without a reason. It is clear that there cannot be *one* answer to this question. The answer that interests us here is the following. The answer takes the situation just described as a starting point; a situation in which one cannot imagine alternatives, and therefore a situation in which the statement 'Here is a table' (after visual and tactile observations) is assumed to be certain. But this situation is interpreted completely differently from the way in which we have interpreted it. Here, the fact that other circumstances are not taken into consideration is not traced back to a lack of knowledge and to a lack of power of imagination, but it is assumed that this exclusion *amounts to a definition* according to which tables only have sensual effects, and no others. Thus, it is assumed that the ordinary person, who rejects the photographic plate because he does not know the process of photography (perhaps an unrealistic example), *does this on the grounds of an implicit decision* not to allow evidence other than sense impressions. This assumption is obviously refuted by the confusion which the 'ordinary person' (if there is such a thing) has when confronted with unusual situations. Let us teach him the process of photography! Let us train him to be an expert photographer! And let us arrange a systematic hallucination of his visual senses and of his tactile senses whose cause he cannot discover on the plate! Will he then triumphantly reply that he has already decided the case long before, and that of course a table is present here? I hardly believe so. He will be confused, and it will require long consideration before he will be clear about what the best procedure will be in this case. Yet this means the best convention, if certainty is in play. Thus, I repeat that the certainty we feel in everyday life with reference to a particular result is a purely subjective phenomenon, and only leads to logical certainty if we install the appropriate conventions. This will become much clearer with the following example.

This example, which I owe to Professor Tranekjær-Rasmussen of the University of Copenhagen,[2] shows that there are statements that are subjectively completely certain in a particular observational situation (or at least as certain as the statement 'I now feel pain' when one is in pain), and which contain a contradiction – which is surely a good reason to doubt their truth. Here, we are not interested in the details of the experiment, but only in the result. Subjects were asked to compare the lengths of three lines: a, b, and c. The result of direct observation (whose absurdity, for the most part, only subsequently appears to the subjects, who are occupied with the correct description of what is observed) is that $a = b$; $b = c$; but $a > c$.

---

[2] [See E. Rubin, 'Visual Figues Apparently Incompatible with Geometry', *Acta Psychologica*, vol. 7, 1950, and E. Tranekjær-Rasmussen, 'On Perspectoid Distances', *Acta Psychologica*, vol. 11, 1955. (Ed.)]

I take this result as philosophically highly significant because it means nothing other than that a statement which seems to fulfil all the criteria of a sense-datum statement (it is 'absolutely certain', its truth is 'immediately given', it describes what is 'immediately present', etc.) can contain a contradiction. And as we always try to avoid contradictions, we seem to be forced to concede that our statement is just not as indubitable as we initially thought.

Against this interpretation of Tranekjær-Rasmussen's result, Professor A. J. Ayer raised the following objection in a discussion with the author: it is wrong to say that the correct description of the impression runs '$a = b$; $b = c$; $a > c$'. Instead, the correct description is that 'it seems that $a = b$; it seems that $b = c$; it seems that $a > c$', and this statement no longer contains a contradiction. Yet, this solution does not work. What I directly observe is not that a seems equal to b. The impression is not indefinite and uncertain. I observe that $a = b$. The element 'seems' does not appear *in* the perception, but only serves to hint that the following report concerns a perception and not a physical object. Thus, the situation can be grasped with a single glance, so 'seems' belongs to the beginning of the description and is equivalent to 'I perceive that', and that is what we have claimed – the existence of a direct description of a perception which contains a contradiction. I cannot help ending the discussion of this example with the remark that in my opinion the theory of sense-data only arises because most philosophers, aside from a certain sloppiness in logical matters, do not know much about the psychology of perception. I believe that a basic knowledge of the psychology of perception must lead, sooner or later, to the insight that the idea of the indubitability of sensations is not only a logical, but above all, a psychological myth.

But let us now finally turn to the logical side of the theory being criticized! That is, let us formulate the conventions which must be made if an empirical statement is to be an observation statement in the sense of the second explanation; i.e., a statement which is indubitable and absolutely true under particular circumstances. And let us also investigate the consequences which such conventions must have.

In order to express these conventions correctly, and also in order to see their consequences as clearly as possible, we first pose the following question: how do we assure ourselves that in the case of observation statements in the sense of the second explanation, we have correctly described the present, 'immediately given' entity if we apply the word 'pain' to it, to take a concrete example? The answer to this question seems trivial. First we identify what is present as pain, and then we apply the appropriate word, namely 'pain'. But how does this identification take place? In the case of a physical object, for instance a chemical substance, the answer is easy: we observe certain circumstances (behaviour in

reactions), we notice that the presence of these circumstances is characteristic of barium sulphate, and we therefore label the substance with the
name 'barium sulphate'. Naturally, it is true that in this case we can never
exclude error. But we always have the possibility of setting up more tests in
order to appease each particular doubt, at least provisionally. How is it
then in the case of pain, assuming that pains are sense-data? If a statement
about pain is to be absolutely correct, then it cannot concern anything that
is not immediately present at the moment of observation. Thus, it is now
no longer possible to introduce various characteristic features for the
presence of pain and, in addition, to distinguish the presence of these
characteristic features from the presence of *pain itself*. Characteristic features
and the object must all coincide in one. But how about the criterion for the
truth of the statement 'I feel pain' in this case? That is, how do we assure
ourselves of the correctness of this assertion at the moment it is produced
(after all, no other moment is in question)? The answer that 'I feel pain' is
correct if pain is present is no longer satisfactory because our problem
consists just in the following: how do we identify pain independently of the
fact that we want to say (that we have strong psychological pressure to say)
'I feel pain'? However, the statement and its truth conditions seem to be
two different things, and therefore it should also be possible to determine
the presence of the truth conditions, i.e., the pain, independently of the
presence of the assertion. Once again, let us consider the case of a physical
object! Here, we also produce a description like 'There is a table in front of
me' because we feel intuitively certain of the correctness of this statement
on the grounds of the observations made, as well as on the grounds of
acquired habits. But in the case of a physical object, this intuitive certainty
is not a truth criterion. Neither is the production of (or the silent
consideration) 'Here is a table' the only criterion for the identification of
tables. The truth of the statement, as well as the identification of the
supposed object, depends on many other circumstances which do not need
to be of a verbal nature; for example, indentations in the carpet, or objects
changing their path (a ball must bounce off a table), and the like. The
identification of a physical object is therefore a process which can be
executed independently of the fact that the correct or intuitively plausible
description is 'Here is a table.' This, of course, has the consequence that
there is a whole series of statements whose truth is decisive for the truth of
'Here is a table', although they do not describe tables but completely
different circumstances. In addition, it has the consequence that 'Here is a
table' is a hypothesis whose future refutation cannot be ruled out. Yet, a
sense-datum statement should be irrefutable. Therefore, it must be contested whether such a statement can concern circumstances which trans-
cend what happens in the moment of its production or of its consideration.
The *only* criterion for the presence of a sense-datum is therefore the

intuitive pressure to produce a certain description or, even better, to produce this description upon query (in other words, introspection results in our simply feeling pain; on being questioned we say 'I feel pain', and no other element appears. That is to say, there is no experience of evidence with reference to the correctness of the assertion obtained in this way.) Or, assuming that pains are sense-data, the only criterion for the presence of pain is the fact that there is a disposition, or a psychological pressure, to say 'I feel pain' (we are, of course, speaking here about individuals who have the further disposition always to tell the truth). Pains are present when, and only when, I feel the urge to say 'I feel pain.' (This urge does not have to be conscious, and it is also not normally so – see the remark in the last parenthesis. It only makes itself noticed in the form of a disposition to a certain action.) This is the convention which we will lay down as a foundation for the discussion to follow.

First, let us pay attention to the fact that this convention is only non-empty if there are statements in the language being used whose correctness appears psychologically evident in a particular observational situation. A very prudent or a very poorly trained observer, who is almost always in doubt about what he should say, could thus hardly apply the convention made. The existence of well-trained observers is an empirical fact. The applicability of our convention is, along with it, equally an empirical fact.

The second remark relates to the fact that a sense-datum cannot be distinguished from the way it is described. As we have said, the only (necessary and sufficient) criterion for the presence of a sense-datum is to be sought in the intuitive certainty of a particular description. As has been elaborated above, this convention must not be understood in the sense that there are now *three* things; namely, the sense-datum, the description, and the fact that this description seems intuitively certain – and that all three of these things can be distinguished from each other in the observational situation. This description of the situation would not only be phenomenologically wrong (if we say 'I am in pain', the pains are the only phenomenon that we can clearly discern), it also has other undesirable consequences. Because, in this case, it seems to be necessary not just to assure oneself of the pain, but also of the presence of two other elements; namely, of the experience of evidence as well as of the statement, if one wants to determine the truth of 'I feel pain'. This obviously leads to an infinite regress. (Also the experience of evidence is certain if and only if a description concerning it is itself evident, and so on.) And second, the distinction between pain and the experience of evidence contradicts the criterion introduced above according to which the certainty with which the statement 'I feel pain' is produced is the only necessary and sufficient criterion for the presence of pain. Therefore, there are not three separate entities: a statement, an experience of evidence, *and the pain*. There is only the process of producing the statement with certainty *and this is already the sense-datum. Hence, sense-data cannot be separated from the process of their description.* This peculiar property of sense-datum statements has already been stated by Plato in the *Cratylus* as a critique of their use. Hegel (*The Phenomenology of Spirit*) has described this property very dramatically, though perhaps not very clearly. We owe an extremely clear analysis of the situation to Schlick (in his article 'Über das Fundament der Erkenntnis', *Erkenntnis*, vol. 3, 1934).

The third remark concerns the fact that not a single known sensation is a sense-datum in the sense just illustrated. We justify this remark with the help of the following example: assume that an individual P dreams that he has the sensation S, and that while dreaming he formulates the statement 'I feel S' so that he can clearly be heard by everyone. According to the criterion just posed, it is already (absolutely) assured that he has the sensation S. Yet, we want to draw a general distinction between the fact that P feels S, and the fact that P dreams S, and this for *every* sensation S. This means that no sensation can be a sense-datum in the sense in which we circumscribed this word in our conventions. Even those statements of our language which we grant maximum certainty, namely statements about sensations, are thus not observation statements in the sense of the second explanation. Yet neither the non-existence of sense-data, nor the impossibility of a language which contains this kind of observation statement is proven by this. It only shows that we must fundamentally alter our present means of communication if we want to talk about our sense-data. The next, fourth, and decisive remark is based on the analysis of the language in which we really can talk about sense-data.

For the purpose of this analysis, let us assume that individual P was instructed to say 'I feel pain' from childhood on when in the presence of the fragrance of cologne. Such training will obviously mean that for P the production of the statement 'I feel pain' when cologne is smelled is evident, and therefore the statement will be a correct, indubitable, absolutely true statement for P according to the convention stated above. Here one might object that for P, 'I feel pain' means something different from what it means for us, and that P 'really means' 'I smell cologne'. Yet, for a sense-datum theorist, this objection is by no means allowed. He has claimed that the *only* criterion for the presence of a sense-datum is the experience of evidence whose description is delivered by this datum. Therefore, he cannot act as if *he*, as an external observer, could determine what is experienced by P. This, of course, has the consequence that he does not know what P is talking about. But even P finds himself in this undesirable situation concerning those assertions which he has expressed in the past. Thus, statements that are observation statements in the sense of the second explanation are, for the most part, meaningless, and they are meaningful only in isolated moments and even then only for a few individuals who can never communicate to each other what they discern in this moment. It is clear that such statements cannot be used to describe the test-conditions of a scientific theory, or even of an ordinary assertion about tables and chairs. Firstly, they can never be published, and thus they cannot be exposed to more tests ('An ascertainment', [Konstatierung] Schlick says, 'cannot be written down'). And secondly, systematic experimentation is completely impossible if the problem which is being decided by the experiment is formulated in statements which nobody really understands, except if one has coincidentally had the 'right' sensation. One should pay attention to the exact nature of this argument. It by no means relies on the extremely persuasive force of the linguistic proof given by Wittgenstein, which shows that sensations, in the sense spoken of in ordinary language, are not sense-data. After all, such a proof can always be countered with the suggestion that this does not establish the refutation of the existence of sense-data, but only the impossibility of describing them in ordinary language. An epistemologist who analyses ordinary language and wants to make it, as well as scientific language, understandable on the basis of sense-data, will just demand that an artificial language which contains observation state-ments in the sense of the second definition must be constructed. Our analysis now shows that such a language can hardly be a possible candidate for a means of communication, and surely cannot be a means for the description of the results of observation. The attempt to anchor scientific theories with sense-data does not lead to the clarification of science, nor to a very certain justification. It leads to its total dissolution. Therefore, if we want to continue science as an inter-subjective and systematic enterprise, then we must no longer associate it with sense-data, and we must do this totally regardless of the kind of language we are using at the moment.

It is very important to notice on what the eliminability of sense-data relies. If one starts with the assumption that sense-data exist independently of our linguistic decisions, then it is really very difficult to see how we can eliminate them with epistemological investigations. But if, as we have attempted to show above, the certainty of sense-datum statements is based on a decision, then the elimination of sense-data is thoroughly at our disposal: we need only retract this decision.

Therefore, our elimination of sense-data is based both on a *decision*, namely the decision to use only such means for describing which enable systematic experimentation and inter-subjective communication of the results obtained by such experiments, as well as on the *knowledge* that a sense-datum language does not fulfil this criterion. The *realization* of our decision is made possible by the *empirical fact* that our inner-life, as well as the inner-lives of others, has certain regularities, that there exist similarities, in short, that those measuring instruments that we call 'human beings' react in a law-like way to their surroundings. Therefore, the decision made is in no way an unachievable ideal. On the contrary, it is far more realistic than the (only rarely explicitly formulated) conventions on which the theory of sense-data is based. The result of this decision is that we do not allow in our language statements which are observable in the sense of the second explanation. And with that, we finally arrive, after a long detour, back at our starting point. We have shown that the theory of sense-data cannot rescue the problem of theoretical entities in its second form from absurdity, because methodological considerations demand the elimination of sense-datum statements.

4 THE STABILITY THESIS: SOLUTION OF THE PROBLEM

Until now, we have investigated the problem of the existence of theoretical entities in two different formulations, both of which depend on the sense in which we use the word 'observable' (on the assumptions which we made about the role of observation). In the first formulation, an observation statement was a statement of which we could quickly and easily arrive at a hypothesis about its correctness. It has turned out that according to this formulation every statement is potentially an observation statement. According to the second formulation, an observation statement was a statement which could be definitively verified on the basis of present data. It has turned out that according to the second formulation every statement must be counted as a theoretical statement. Therefore, both formulations lead to the breakdown of the problem of the existence of theoretical entities which, after all, just consists in the question about the reasons one has for the assumption of the existence of theoretical entities *presuming* that the existence of observable entities is unproblematic. Now, is there an explanation of the concept 'observable' which is intuitively cogent, which corresponds to the practice of the process of scientific observation, and which, in addition, is constructed in such a way that not every concept is either an observational concept or a theoretical concept? That there is such an explanation is claimed by an argument used by Professor Feigl in discussions with the author.

In order to develop this argument, we need only allude to the fact that observation statements normally have the function of deciding between rival theories. Thus for example, one draws on an observation in order to determine whether the wave theory or the particle theory of light is correct (Foucault's experiment). If, the argument continues, an observation statement is to decide between two alternative theories, then it must be an impartial judge. In particular, its *meaning* must not depend on the meaning of the descriptive concepts of either theory. Because this applies to every pair of theories, it follows that there must be statements whose meaning is independent of the structure of every conceivable physical theory. These are the observation statements. Therefore, observation statements are statements which can be explained without reference to theories and whose meaning is also independent of changes in the 'theoretical superstructure'. However our theories may be constructed, the assertion 'pointer A coincides with graduation mark n' has one and the same meaning, and is therefore invariant with respect to change of theories. In an earlier treatise,[3] I called the claim that the meaning of observation statements is independent of change of theories the *stability thesis*. One can also say (*a third explanation*) that observation statements are statements which obey the stability thesis.

It is easy to see that the argument which has just been developed has a hole. Just from the fact that the decision between two theories A and B demands a statement c whose meaning is dependent neither on A nor on B, it does not follow that the meaning of c is independent of *every* theory. It only follows that the meaning of c cannot be dependent on A or B. It can still be determined by a theory $C \neq A$; $C \neq B$, where neither A nor B need be a rival to C (example: the observation of the deflection of light at the edge of the sun, A = Newton's theory, B = the general theory of relativity, C = the particle theory of light). The conclusion that the meaning of an observation statement cannot depend on *any* theory requires another premise, and this premise is usually stated as the *postulate of the homogeneity of experience*. Let S be a statement which plays the role of an observation verdict in the decision between A and B. Then S, or the negation of S, must be capable in principle of playing this role with reference to any other pair of theories. The content of this very plausible postulate states that in principle it must be possible to test every theory in every domain of experience. Yet by this, nothing is said about the meaning of S. There is *one* experience of red, but a statement which is expressed on the occasion of such an experience could be a statement about sense-data, a statement about the colour of physical objects, and the like. Therefore, we must make the additional presupposition that an observation statement, for instance the statement S, is not only connected to the same experience in all contexts, but in addition has the same meaning. This is the thesis that has to be proven. The only argument for the stability thesis which is persuasive, in my opinion, is the following argument which Bertrand Russell has developed in his book *Inquiry into Meaning and Truth*.[4] When we test theories with observation statements whose meaning depends on other theories, we are forced to accept a coherence theory of truth. We do not reject a theory

---

[3] [The reference is to material which became 'An Attempt at a Realistic Interpretation of Experience' (1958b) now reprinted in *PP1*. (Ed.)]

[4] [(London: Allen and Unwin, 1940). Feyerabend is referring to chapter 10, especially p. 139. (Ed.)]

because it contradicts the facts. We reject it because it contradicts a certain theory on which we base the construction of our observation language. A coherence theory is untenable. It is missing reference to facts, and thus, and this is Russell's argument, there must be a language which does not depend on any theory, and this is the observation language.

We will analyse Russell's argument later in this section. We will prepare for this analysis by investigating the properties of a language which suffices for the stability thesis. Let us assume that there is such a language. It will be a language in which we describe certain things, in which we ascribe properties or relations to them. Naturally, a case could occur in which one or another of our descriptions is false because we have not proceeded with sufficient care. But could it turn out that the categorical system of the observation language is wrong, i.e., could it turn out that even adequate and accurately executed measurements lead to incorrect results? In order to answer this question, let us look at the situation in a *theory*, for instance the classical atomic theory. This theory works with certain basic concepts (atom, force, distance, etc.), and it describes certain events, such as the fluctuation of pressure in a vacuum, with the help of these basic concepts. As it is a theory, it is vulnerable to refutation. And if it is refuted (in a decisive experiment with a continuum theory), then we must admit that everything was mistaken, that there are no atoms, and therefore that certain concepts must be removed. (As we know, one cannot describe existing objects based on concepts which refer to non-existing objects.) The fictional observation language with which we are dealing at the moment is allegedly free of theoretical elements, and in addition it is stable (for the reason just given, in fact). Therefore, its categorical apparatus must always be adequate, and we could never arrive at a situation which forces us to reform this apparatus. This has a very important consequence. Let us assume, for a moment, that Carnap's 'thing-language' [Dingsprache] is an observation language in the sense just discussed. (The 'thing-language' is a language in which observable properties, like a certain colour or form, are ascribed to macroscopic objects like tables and chairs.)[5] Then, after a well-defined series of observations, it is impossible to doubt the existence of tables and chairs. Their existence is now a fact which is absolutely fixed. Take note, the last statement is not a tautology! If we *presuppose* the existence of phlogiston and, in addition, if we make certain assumptions about the properties of phlogiston, then we can make a clear distinction between adequate and inadequate methods of observation and measurement of

---

[5] [In Carnap's 'Die physikalische Sprache als Universalsprache der Wissenschaft', *Erkenntnis*, vol. 2, 1932 (English translation: *The Unity of Science* (London: Routledge, 1934)) the terms 'physical language', and 'physicalistic language' are introduced. Carnap himself uses the term 'thing-language' (as an abbreviation for 'physical thing-language') in 'Logical Foundations of the Unity of Science', *Encyclopaedia of Unified Science*, vol. 1, 1938. (Ed.)]

phlogiston on the grounds of these presuppositions. If we *presuppose* the existence of atoms and if, in addition, we again make certain assumptions about their properties then, again, we can make a distinction between adequate and inadequate methods of measurement of their diameters or the number of atoms (molecules) in a molecule.[6] If, however, there are no atoms, then the result of all those measurements which were thought to be adequate within atomic theory requires a re-interpretation. The number $\sim 10^{-8}$ no longer refers to the minimal extension of matter, but may refer to a basic constant of a certain periodicity of the continuous medium. The stability thesis leads to the result that in the case of an observation language, a re-interpretation of this kind is neither necessary nor possible. This means that if there is an observation language in the sense presently discussed, then certain observations are totally sufficient, by themselves and without any further theoretical assumptions, *to claim with absolute certainty* the presence of a particular state of affairs (see also the consideration in the previous section). A philosopher or a scientist, who accepts the principle that every statement of science must be revisable, must therefore reject the stability thesis, and with it the existence of an observation language in the sense of the third explanation. (Whether this means the acceptance of a coherence theory of truth, as Russell claims, will be investigated later in this section.) One should notice that this rejection is also grounded in the methodological *decision* to incorporate only refutable statements into science (or into our knowledge generally). (According to my view, this is an entirely general property of epistemological problems. They are not solved by *proofs*, but by *decisions*, as well as by the (empirical or logical) evidence that the decisions made are realizable. So, for instance, science, in the sense of certain methodological conventions, is only realizable if there are individuals whose language has, besides an emotional, descriptive, and indicative function, a well-developed argumentative function as well; and if, in addition, these individuals are capable of rejecting plausible ideas and accepting or inventing implausible ideas. The role of decisions in the discussion of epistemological problems has been emphasized with great clarity by Professor Victor Kraft as well as by Professor K. R. Popper. The question of the realizability of such decisions, on the other hand, has not yet received the attention that it deserves.)

We now turn to the question of the existence of languages which suffice for the stability thesis. Philosophers of the most different origins have claimed that *ordinary language* or the *thing-language* (which is a part of ordinary language) satisfies the stability thesis. This claim must be investigated with great care. It is entirely possible, and maybe even right, that ordinary language is stable with respect to the change of scientific theories. But from

---

[6] ['mole' (Trans.)].

that, it neither follows that it does not contain theoretical elements, nor that its categorical apparatus is adequate for describing reality. In addition, we can show that a language which not only suffices for the stability thesis *de facto*, but must suffice for logical (or 'ontological') reasons, cannot play a role in testing scientific theories, and therefore cannot be an observation language. This is because a necessary condition which every observation statement must fulfil is that it should be derivable from the theory it should test. We will now discuss the last two claims in order.

First, the *de facto* stability of ordinary language, if it exists, shows at most *that nobody has executed a change*. It does not show either that no change is necessary, or that such a necessity will never occur. Thus, it does not show that ordinary language does not contain theoretical elements. After all, a language does not change by itself. It is a product of the human beings who speak it and, therefore, it reflects the ideas, the views, and also the behaviour of those human beings. The reason for the stability of the language therefore could simply be laziness, or ignorance, or dogmatism (the latter either calls on 'the tradition' or on 'ontological' pseudo-proofs which all contain the *non-sequitur* – what is so, must be so). Therefore, the *de facto* stability of a particular language should not impress us. Instead, we have to pose the *logical* question of whether a revision is *imaginable*. But this question is identical to the question of whether ordinary language contains theoretical elements.

A very simple analysis shows that the question has to be answered 'yes'. Let us consider, for instance, the way in which the word-pair 'up-down' is used by human beings who do not know anything about the structure of the world (this is not a merely hypothetical case, and if one were to object that we are now discussing a language that is no longer spoken, then one would have already admitted that the stability thesis, *de facto*, is wrong). 'Up' means, in this case, the direction from the toes to the head. 'Down' is the opposite direction for an erect individual. One sees that this method of application is absolute from the popular objection which was made against the assumption that the earth is round, i.e., that the antipodes must fall 'down'. Mark you, this empirical claim is the consequence of the *method of application* of the word-pair 'up-down', and not only a consequence of a physical theory alone. One always falls 'down' (a generalization from experience). 'Up-down' is an absolute direction in the universe. *Therefore*, if there are antipodes, they must fall off the earth.

Let us consider this method of application more closely. It implies a very interesting cosmological theory. According to this theory, the universe is anisotropic and possesses a distinguished direction. Thus, ordinary language is far from being an observation language in the sense of the third explanation. It contains theoretical elements. These theoretical elements are very abstract, and they go far beyond what is directly observable. (That this is generally valid for 'ordinary languages' has been shown by Whorf in his superb investigations.)[7] In addition, these elements are constructed in a way which contradicts modern knowledge. This has indeed been acknowledged, and therefore today we use concepts like 'up' and 'down' in a *relative* sense in which we relate them to the centre of the earth or, even better, to the strongest neighbouring source of force. (The appearance of centrifugal forces in airplanes leads to more complications and more relativizations.) Thus, instead of an absolute predicate which so to speak describes an inherent property of space, today we have a relation. Of course it is perfectly possible that certain language-groups have not executed this change, out of ignorance or dogmatism. As we have shown, this still does not make ordinary language an observation language in the sense of the third explanation. But we must admit that such behaviour is still entirely possible. Yet, the consequence is that ordinary language can no longer be a candidate for an observation language for those theories which contradict their theoretical construction. This is the second objection which we stated above.

The objection is as follows. An observation language must fulfil two conditions. First, it must be constructed in a way in which human individuals can accept or reject any of its singular statements quickly and with certainty. (In the *first explanation*, this property played the role of a necessary *and sufficient* condition. It is clear that all further explanations must contain this property as a *necessary* condition.) But, second, the statements of an observation language must be derivable from the theory to be tested. What now follows from the theory of gravitation (which shows the inadequacy of the old usage of language) are statements about the relative positions of mass-points or planets, thus statements which require a *new* meaning of 'up' and 'down' for their formulation. These statements cannot be expressed in the old language, and with that, it ceases to be an observation language, at least with respect to this theory. But this result is already sufficient for the refutation of the stability thesis, which claims that ordinary language is an observation language *for every conceivable theory.*

A situation which wrongly suggests the correctness of the stability thesis, even in those cases in which a very revolutionary change in word usage has occurred, is the following. If we restrict ourselves to the events near the surface of the earth (and these are the only events which were of interest to the ordinary person, at least until around 1920), then the old word usage coincides with the new one. One can only arbitrarily restrict 'ordinary language' to this domain, and then announce that no change in word usage has taken place. But, firstly, this restriction means a modification, as the

---

[7] [The reference here is to papers reprinted in Benjamin Lee Whorf's *Language, Thought, and Reality* (Cambridge, MA: MIT Press, 1956). (Ed.)]

argument against the possibility of antipodes, which was formulated *in ordinary language*, shows. And, secondly, two methods of application are still not identical, even if they coincide in a narrow domain.

A second example that shows the necessity of a change in ordinary language is the following: in everyday life, we directly assign the colour we observe to the observed object. That is, a statement which we express on the basis of certain impressions describes neither our own sensation, nor the atmosphere between the object and the observer. The statement concerns the observed object itself, and it ascribes to it a property which is assumed (on the basis of the categorical system of ordinary language) to be independent of how it is illuminated, of the state of the observer, as well as of the actions which the observer is just executing. Therefore, this property is an objective feature of the observed body. The discovery of the Doppler effect forces us to the following modification: what we assign to the object on the basis of 'direct observation', i.e., on the basis of just those actions which earlier led us to assign objective qualities, is no longer an objective or absolute property of the object itself, but a relationship between it and the co-ordinate system in which the observer is at rest. A change in the concept 'colour' has proven to be necessary. We can accomplish this change in a two-fold way: either we introduce a property which is no longer accessible to direct observation, and which we call the 'eigen-colour' of the object, or we can continue to interpret the word 'colour' as an observation term, in which case we must re-interpret it as a relationship, and no longer as an absolute property of the observed object. It can also be shown here that ordinary language can only be a satisfactory observation language for optical theories if one has carried out these modifications.

At this point, the following decisive objection arises: why do we demand that an observation language for optics must conceive of the colour observed as a relation, and no longer as an objective property? Because of the Doppler effect, i.e., because a part of optics teaches us that the properties of a wave-train which enters our eye depend not only on the vibrations of the electrons at the surface of the observed body (here thinking classically, and also speaking only about self-illuminating bodies), but also on the relative motion between the receiver and the sender. The same, and this was our argument, must therefore apply for the observed colour. Yet this argument, according to the objection, presupposes that the problem of the existence of theoretical entities has already been solved in the positive sense. After all, we do demand that the observed colour will be interpreted as a relation because *in the theory* the properties of the arriving wave-train, for instance its wavelength, depend on the co-ordinate system from which it is observed. Clearly, by that we assume that there *are* wave-trains, that their properties are objective features of nature which therefore must come into play in the case of observations. Admittedly, this objection will continue, a realistic interpretation of physical theories must force us to change even our observation language. But the question consists exactly in whether such a realistic interpretation is justifiable, and this question cannot be solved by a discussion of the circumstances which occur if we accept realism. But if, on the contrary, we assign no meaning whatsoever to the symbols of a theory, if we only interpret them as an adequate instrument for predictions of known observable facts which can therefore be described in observation language, then there is not the slightest reason to change the categorical system of this language. With this, we are once again back at our problem, i.e., of the question of whether or not theoretical entities should be interpreted realistically.

Yet, this question now stands in an entirely new light. We have proven that ordinary language, or the thing-language, which should here be defended, contains abstract theoretical elements. How are these theoretical elements distinguished from the ideas which certain scientific theories have as their foundations? They are only distinguished by the fact that they appear in ordinary language and are used for the description of observable states of affairs. That they appear in ordinary language is a historical coincidence which has nothing to do with questions of existence or the appropriateness of these elements. We can say the following about the fact that they are used in the description of observable states of affairs: with two examples, we have just shown that experiences are describable not only with the help of the categories of ordinary language, *but just as well with the help of another categorical system* which in addition has the advantage of greater coherence. Therefore, what remains as an argument for the distinguished position of theoretical assumptions which have the thing-language or ordinary language as their basis? Only the remark that these assumptions are 'intuitively evident', which means nothing other than that we are more acquainted with these elements, and less acquainted with other elements. And this also has nothing to do with questions of existence or with questions of truth. If it is admitted that the intuitive reasonableness of speaking in a certain way is a criterion for existence, then we only need to retrain the observers in order to convert them to our point of view.

From all these considerations, it follows that the arguments which are usually stated for the choice of a certain observation language (in the sense of the third explanation) are totally insufficient. Every theory which has sufficient descriptive diversity, and in addition an approximately correct description of events in the everyday world, every such theory has an equal right to claim that it provides the required categorical system for an observation language. The difference is, of course, that some of those theories will be wrong (which is proven by investigations *outside* the everyday domain). With that, the thesis that realism about the observable world is self-evident, while realism about theoretical entities first needs a

THE PROBLEM OF THEORETICAL ENTITIES

justification, collapses in on itself. The only problem that remains is the question of whether something exists at all. And if we admit that observable things exist, then it follows completely from that alone that theoretical entities also exist, because observable things are the result of the over-lapping of theoretical entities, and a composition of nothings cannot lead to a something.

This, then, is the solution that we suggest for the problem of the existence of theoretical entities: every observation language contains theoretical elements (this already follows from our rejection of sense-data). Those philosophers who pose the problem of theoretical entities grant that observable things exist. The categories which we use for the description of observable situations are not yet unequivocally determined by the structure of experience which, as we know, can always be subject to errors, and which, in addition, only provide approximately true information. Thus, it is possible, and also indeed the case, that completely different theories adequately describe the same everyday experiences with the help of totally different categories. The principle that what is observable in everyday experience is also real leads immediately to the declaration that the elements of *all* these theories are real, unless we possess a principle of selection which transcends everyday experience. This principle of selection is the scientific experiment. Experiment selects, in the ideal case, a single theory from the theories which are adequate in the domain of everyday experience. The principle that what is observable is also real leads immediately to the result that the entities with which *this* theory works are themselves real. Briefly, the positive solution of the existence of theoretical entities follows from: (a) the fact that the formulation of this problem silently presupposes the assumption that all observables exist, and (b) the assumption that the observation language is nothing other than a subjec-tively distinguished section of a very general and abstract theoretical system.

Finally, we must answer Russell's argument. The thesis which follows from the arguments we have given above is that every language is a theoretical language, i.e., a language which contains an abstract, detailed, and changeable system of categories, and that the observation language is not an additional language which no longer contains theoretical elements, but rather is the logical sum of all those parts of different theoretical languages now in use, of which human individuals can quickly come to a decision which will be unanimous. And the problem of the existence of theoretical entities is to be solved positively, because it was posed under the presupposition that observable things exist. But where are the referents of a theory? And are we not forced to accept a coherence theory of truth?

As so often in philosophy, the answer is – yes and no. It is true that our 'experience' continually changes, not only extensively, but also concerning the 'nature' of the already observed facts. Statements of lengths used to have an absolute meaning. Today they are re-interpreted as relations which are valid in particular, specified co-ordinate systems. The observed star positions used to have a direct relationship to the stars themselves. Today, we say that we only observe 'directly' the direction of the arriving light, and this is not directly connected with the location of the star emitting the light. Thus, we re-interpret our 'experiences' in light of the theories that we possess. There is no 'neutral' experience. Despite this re-interpretation, there is still a practical invariance in as far as we demand that even the most comprehensive and most abstract theory will give us approximately those results which a less abstract theory contains. This methodological principle has, as a consequence, that all the different theories which have been proposed in the course of the history of science have 'something' in common. What is this 'something'?

This something is by no means a 'fact', and this because, after all, the different theories, including the theories always implicit in the observation language, express something completely different about this common core. In particular, this core is not a sensation because if we interpret our theories realistically, and we have seen that the way in which the problem of the existence of theoretical entities was posed forces us to such an interpretation, then we do not speak about sensations except in the case of a psychological theory. The common element of different theories is therefore nothing which is describable *qua* element in any theory. Of course, one can ask for information from the latest, most modern, successful theory. But even this theory will be overtaken, and we will still have to demand that its successor will be successful in the same domain in which it itself was successful. Therefore, we must admit that the theoretical assertions that we make are posed for reasons of *coherence*. At this point, I would like emphatically to point out that this cannot be an objection. If we possess a correct theory, then it is indeed appropriate to demand that all other theories conform to it. The correctness of a theory admittedly lies in its correspondence to things. But the correspondence does not lie where Russell looked for it; it is not in the impressions of the subject, whether they are interpreted as sense-data or as sensations in the usual sense. The identification of correspondence and impression would mean nothing other than the elevation of the observer to a measuring instrument of absolute precision which reflects exactly what happens in the world. 'What the observer perceives is also real' is nothing but the purest subjectivism. But how do we *ensure* ourselves of this correspondence?

I believe an answer to this question can only be given by a fully developed causal theory of perception. This theory would have to show in what way certain facts in the environment lead to reactions in organisms which have a certain (bodily and spiritual) construction, and how a certain series of reactions can give rise to expectations. It is clear that our expectations, which after all rely on experiences in a narrow domain of the world, will no longer be adequate in a journey to other domains. The totality of all reactions (sensations and perceptions included) in the domain from which one starts are now these common elements of which we talked earlier. Yet, this element is not a *fact on which our theories rely*, but *it is the act of proposing these theories itself*. As observation mechanisms, we cannot avoid reacting in a more or less obvious way in certain situations. As rational beings, we want to interpret these reactions as descriptions of present facts. The disappointment of the expectations to which the categorical system leads, which we have introduced for the purpose of this description, forces us to link the same reactions with other expectations, i.e., it forces us to assign another interpretation to the expressed statements (or the experiences enjoyed). What is common to all theories is therefore nothing that can be described by any of these theories, because it consists simply in the fact that sometimes certain assertions, and among them also the theories mentioned, are produced almost automatically. The 'unity of experience' is therefore a *practical* unity which is based on the unity of the structure of human organisms. This unity does not provide us with a point of reference, and is no foundation of confirmation for our theories, although scientific methodology demands that all theories in the everyday domain lead to predictions which are automatically produced (not confirmed!) by human organisms in this domain. Theoretically put, this naturally means the acceptance of a coherence theory. But the theories used are constructed in a way that particular consequences of the theories are expressed by human individuals under particular circumstances, and therefore become a part of human behaviour – although naturally this has nothing to do with their meaning or their correspondence. The extent of the correspondence of this behaviour can only be decided by a causal theory of experience.

2

Knowledge without foundations

(1) Philosophy today – not least the theory of knowledge – is a very confusing subject. It deals with highly technical and esoteric matters which can be understood only by the initiated. The problems treated seem to have little to do with things that are of common interest, such as the improvement of the life of mankind. Indeed the philosophical problems are sometimes quite explicitly and even proudly declared to be irrelevant to such concerns. Admittedly attempts are made to establish that philosophy is irrelevant to these concerns, but the attempts are often exaggerated – if not incomprehensible. The worst thing, however, the perennial scandal of philosophy, is the quarrel of the schools. There is no problem that is generally accepted, not one point of view that is regarded as binding for all. And it seems to be possible to propagate the greatest absurdities by advertising them as the basic tenets of some new school.

This picture is very different from a certain ideal which we frequently notice in the history of thought. According to this ideal, philosophy reigns supreme among the enterprises of the human race; it provides a guide to the truth or, at least, an infallible measure for distinguishing what is valuable from what is worthless; and it gives meaning to any particular activity, however specialised, by showing how it is connected with a generally accepted purpose. This is a lofty ideal indeed and one that would seem to be very much in need of realization. Yet whenever a philosopher has attempted to work out this ideal in greater detail the final effect has been the creation of just another school and a corresponding increase of the confusion already in existence. From the very beginning philosophy has therefore been regarded with suspicion, and the belief has arisen that its attempt to provide a universal measure of meaning and validity is altogether hopeless and that it had better be given up.

In the lectures to follow I shall try to show that this belief is not true. I shall try to convince you that if one is sufficiently modest, one need not fail and that one will then be able at one stroke to lay the foundations of an ethics and an epistemology that is more than an exercise in sophistication and 'depth'. As opposed to some philosophical theories which have been discussed in the past, the ethics and the theory of knowledge I am talking about are both very simple. Moreover they have been available for quite some time. Like so many other things, they were invented by the Greeks, more especially by the Ionian philosophers of nature.

(2) In about 550 BC the Ionians populated a narrow strip of land in Western Asia Minor. They were settlers who had immigrated from Crete and from the Greek mainland. The harbours which they founded were used by Greek, Phoenician and Egyptian sea-lines and were connected through caravan-roads with the whole of Asia. *Miletos* alone had four harbours, and its wealth was mainly due to very active sea trade. (Milesian pottery of the seventh and sixth centuries has been found in Egypt, Anatolia, and in South Russia.) Some of the Ionian cities sent out colonies to the coast of the Mediterranean, the Sea of Marmora, and the Black Sea. No doubt this activity was partly caused by overpopulation. The rich noblemen who on the mainland had formed a rather closed and privileged agrarian aristocracy now engaged in seafaring: they developed the art of navigation; they increased the knowledge of geography, of other peoples, their customs, their contributions to knowledge, and their practical efficiency. This new activity leads to interesting developments. The emphasis shifts from the city, or the community, to the individual; from strength and bodily superiority to intelligence and eagerness to explore; from innate excellence to excellence on the basis of achievement. Odysseus rather than Achilles is the hero of the age.

A cosmopolitan spirit began to reign in the cities, as well as a spirit of criticism. This spirit of criticism is closely connected with the realization that even the most fundamental social institutions and even the most basic beliefs are of *human origin* and therefore possibly imperfect and in need of improvement. After all, these people had been the *founders* of the new cities; they had been the conscious inventors of laws that took cognisance of the particular surroundings of each city and of the particular circumstances of its foundation. Also they were the contemporaries of some of the most revolutionary developments on the Greek continent – first the gradual replacement of the king by the aristocracy, his devaluation from an all-powerful monarch to a chairman of the council of elders; then, even more revolutionary, the replacement of a law that is administered by an aristocracy and is grounded in their assumed wisdom, by a *written law* that is binding for all and that establishes thereby an objective and impartial code of justice. True, these developments were not received with universal acclaim. They were regarded by some as a sign of dissolution, of the breakdown of an order which alone can confer dignity upon those participating in it (as kings and rich noblemen, that is). But they also created a tremendous optimism, a firm belief in the power of the human intellect to understand the world and to make suggestions for its change, and in the ability of the human will to carry out these changes. It is this background, rather than the 'divine' Plato which we need to bear in mind when discussing the origin of philosophy as a rational enterprise, and when trying to evaluate the achievements of the first philosophers, Thales and his pupil Anaximander.[1]

(3) Two introductory remarks are in order. First, that our information concerning these early thinkers is very scanty. Nothing has come down to us from Thales himself or from his contemporaries. What we know about him is based upon reports by Plato, Aristotle, Diogenes Laertius, Plutarch Simplicius, and others. Plato cannot always be trusted; he 'like[s] making fun of the Presocratics', as Kirk and Raven have put it. Aristotle possesses a fully developed philosophical terminology and in its terms he describes the theories of his predecessors, who did not possess such a terminology and who were perhaps not even interested in developing one. The same is true of the later Peripatetics who wrote on the history of philosophy. Yet what emerges despite all these difficulties is that these early thinkers were very different from most professional philosophers today.

To start with, the distinction between philosophy and other disciplines was not yet drawn. What counted were ideas that advanced knowledge, that threw light upon existing problems rather than philosophical points of view, the latter tending to restrict research to certain aspects of these problems. Philosophy, moreover, was not so full of technicalities as it is today. I know that some of my colleagues regard this as a disadvantage in Greek philosophy, and as a sign of the immaturity of these early thinkers. I hope I shall be able to convince you that things are just the other way round, that the exaggerated use of technicalities and the emphasis on precision rather than on scope has obscured rather than clarified matters, and that it is possible to deal with the most fundamental problems of knowledge in a manner that makes them accessible to all. Bacon was right when he said that 'knowledge, whilst it lies in aphorisms and observations remains in a growing state; but when once fashioned into methods, though it may first be polished, illustrated, and fitted for use, it no longer increases in bulk and substance'.

These early philosophers were also very bold and optimistic. What they attempted to create was nothing less than a theory of the universe as a whole. This boldness and optimism can be matched only by the boldness and optimism of a Descartes, a Leibniz, a Galileo, or an Einstein. With few exceptions our present-day philosophers are much more cautious. They are almost jittery in their concern that too much imagination, too broad a

---

[1] For the history <see> Werner Jaeger, *Paideia: the Ideals of Greek Culture, volume 1* (New York, 1939); G. Sarton, *A History of Science, volume 1* (Cambridge, MA: Harvard University Press, 1952); K. R. Popper, 'Back to the Presocratics', *Proceedings of the Aristotelian Society*, vol. 59, 1958–9 (Now reprinted in Popper's *Conjectures and Refutations: the Growth of Scientific Knowledge* (London: Routledge, 1963). (Ed.)). For texts and comments <see> G. S. Kirk and J. E. Raven, *The Presocratic Philosophers* (Cambridge: Cambridge University Press, 1957).

vision, may lead straight to disaster. This is why they deal with minutiae and put their trust rather in a careful procedure than in one as reckless as that of the Ionians. The late Professor Austin has produced a historical justification for his own brand of microscopic philosophy. He has pointed out that philosophy in the Grand Manner such as has been developed by Plato, Descartes and Kant has been a failure. This remark is not original. Almost every system builder in philosophy starts with a survey of the schools and the comment that so far philosophy has been in utter confusion and that a new beginning is needed. But it does not follow from such an opinion, which is undoubtedly true, that narrow-mindedness is a virtue. For we shall soon show that philosophy has failed (if it has failed, that is) because of its dogmatism *and not* because of the generality and boldness of the statements made by its inventors.

Finally, the early Ionians were highly cultured. They were curious and well informed in almost every domain of human endeavour. I do not intend to say that contemporary philosophers know nothing but philosophy. Yet, however great their non-philosophical knowledge and ability, their idea that philosophical problems are something special and distinct from the rest prevents this knowledge from having any influence upon their philosophy – with the result that their philosophy is dry, untutored, uninformed. Very often we now witness the curious spectacle of intelligent people with wide interests producing a narrow, boring, and utterly irrelevant philosophy. This is very regrettable indeed. It is not enough to send articles to *Mind* and now and then to write a novel or listen to a concert. Unless the ability in the latter field and the knowledge obtained here (if there is any) is directly applied to one's philosophy (and vice versa) it might as well not exist in the same person. There are notable exceptions, to be sure. But the phenomenon described is still so widespread, and the tendency to specialization and corresponding boorishness so strong, that it must be mentioned.

All these positive qualities of the Ionians are exemplified in the most exquisite manner in *Thales*. Here is a man who is interested in almost everything that is going on around him. He actively participates in the political life of his city, Miletos. Sensing the growing power of Persia, he advises the Ionian cities to form a federation stronger than those already in existence, and he urges them to establish a general council in the centrally located Teos. He seems to have failed in this – it was just not possible to make the colonies accept political commitments that would further restrict their freedom. This is a very healthy sign, but the effects of the Persian conquest, which took place when Thales was still alive and which eventually led to the destruction of Miletos, show how delicate is the balance between freedom and that amount of co-ordination and restriction which is necessary to cope with outside (and inside) dangers. Thales also seems to have had talents as an engineer, for according to the common account of the Greeks (which is doubted by Herodotus) he helped Croesus' army across the river Halys by building a channel that divided the river in two, thus diminishing its breadth. He seems to have travelled widely, for there are various reports of a visit to Egypt. One of them relates how he measured the height of a pyramid by measuring the length of its shadow at a time when the shadow of a small vertical stick was equal to its height. He also seems to have acquainted himself with the geometrical procedures of the Egyptians, but seems to have tried to reduce their numerous problems to a few simple principles which could be easily understood. He is even successful as a businessman; for, guessing that there would be a large olive crop 'he raised a little money while it was still winter and paid deposits on all the olive presses in Miletos and Chios, hiring them cheaply because no one bid against him. When the appropriate time came there was a sudden rush of requests for the presses; he then hired them out on his own terms and so made a large profit' (Aristotle's report). He developed simple ideas about the structure of the earth, about meteorological phenomena and magnets. And he is the first thinker to suggest a theory of matter, as we would call it today. It is the analysis of this theory which will lead us right into the centre of the theory of knowledge.

However, one precautionary remark should be made before proceeding in this direction. As I said above, the reports about Thales and his immediate successors are very scarce and come from much later sources. It is therefore always possible to object to the picture just given by pointing out its very speculative character. Yet even if Thales had never lived, even then the reports are reports of what was at the time thought to be humanly possible; they are reports about a certain form of human existence. It is this form of human existence in which we are interested and which we intend to introduce by discussing Thales. It therefore does not really matter whether there was a person who actually did what Thales is said to have done.

(4) Thales' theory of matter can be presented in two steps. The first is that the various materials (wood, metal, earth, fire, etc.) which exist on the surface of the earth are all basically one and the same substance: there is only one substance, one basic stuff. Variety exists because this basic substance can appear in different states or modifications. This assumption of a unity behind the variety has encouraged the development of the natural sciences ever since it was first introduced. Modifications did occur, but whenever these modifications led to too complicated a substructure of material objects, the attempt was again made to reduce the new variety to an even more fundamental single substance. Heisenberg's non-linear field theory which attempts to explain the thirty-odd elementary particles of today on the basis of some single and all-pervading substance is the most

recent step in this development, a step which, as Heisenberg explicitly admits, was inspired by the ideas of Anaximander, the pupil of Thales.

According to Thales' second assumption, the basic substance is to be identified with water. Water is known to occur in various forms. For an inhabitant of Asia Minor, especially of the coast, it is almost omnipresent. It is necessary for the support of life. Hence it must be present even in those places where there is no water as we usually know it, although there are still lower forms of life. Earth, or sand, then, must be capable of having some of the functions of water; it must be a modification of water. Some such arguments may have led to the identification of the basic substance with the already known element water.

(5) This very simple and almost naïve theory has many interesting and surprising aspects. Discussing it will lead us into the very centre of the theory of knowledge. It will emerge that Thales has not only succeeded in making relatively new suggestions concerning the world we live in, but that his *attitude* towards these suggestions, and indeed towards any theory, reveals a new point of view concerning the nature of knowledge and the means of its improvement. It will also emerge that this point of view is in some respects very similar to the point of view adopted by the great cosmologists of all ages up to and including Einstein. This means, of course, that the discussion of Thales' theory will enable us in a very simple fashion to evaluate the properties and the merits of what is nowadays called the scientific method. This method, far from being restricted to the elaboration of laboratory-matters, involves an attitude of criticism that is applicable in all domains of human life, scientific or not, and that is radically different from the attitude connected with many political, religious and philosophical doctrines of today. I shall try to show what the differences are, and I shall in this way try to convince you of the superiority of the critical attitude of the Ionians. However, before doing this I would like to draw your attention to some features of Thales' theory which are shared both by scientific theories and by mythological schemes of explanation, whose character is not too well understood by some contemporary defenders of what they call the 'scientific method'.

(6) To start with, the theory is *general;* that is, it applies to every object and not only to a selected group of objects. It is *explanatory;* that is, it not only gives a *summary* of the properties of the objects considered, but also a *rationale* for the occurrence of these properties and for their change. The rationale is that since water is the basic substance, a substance plainly essential for life, it is to be expected that things will appear in different forms and that the world will be favourable for the development of organisms. The theory is *counterintuitive* for any person who believes in the existence of a variety of substances and who has cultivated this belief for such a long time that it has now become evident for him. It is *counterinductive*, too; that is, it contradicts

even the evidence of the senses, for we actually experience a tremendous variety of objects, substances, processes. The assertion that all this is basically one is not only not supported by our sensations, it is contradicted by them: just *look* at the difference between metals and, say, the air.

In this respect our theory is very different indeed from empirical generalizations such as 'metals expand when heated'. According to a belief which is fairly widespread, a statement of this kind is obtained in two steps, namely (a) collection of relevant observations O (which in our case describe the behaviour of individual pieces of metal under varying temperatures); and (b) transition, 'by induction', from O to a general statement, G, which is in our case the statement that metals expand when heated. I do not intend to discuss in the present paper all the futile attempts that have been made to justify the transition from O to G. All I want to point out is that the transition to be justified is one from an assertion O expressing observational knowledge to a more pretentious assertion which repeats O *but which also goes beyond it*. It is this 'step beyond' which makes general statements dubious for a strict and cautious empiricist and which, according to him, is in need of justification.

Now, for a cautious empiricist of this kind, Thales' theory would seem to be even worse, *since it does not even repeat the evidence from which it might have started*. It flatly contradicts this evidence. It asserts actual unity where there is observed variety, the presence of water when some quite different material is observed. It not only goes *beyond* experience, but also *against* experience. It would therefore seem to be a paradigm-case of undisciplined speculation, that alleged arch-enemy of Good Science. The interesting thing, however, is that some scientific theories, too, far from repeating the evidence existing prior to their invention, actually go against this evidence. This is far from being a disadvantage. A theory that is inconsistent with experience – and therefore <gives> an account of the world that is different from what is suggested by the sense<s> – provides means for investigating not only certain speculative schemes but even observation reports, which are, after all, not absolutely trustworthy. Such a theory is therefore a much better instrument of criticism and improvement than is a theory that leaves everything as it is on the observational plane. Viewed in this way some contemporary theories which pride themselves <on> their complete agreement with experimental results must be regarded as definitely inferior to Thales' 'naïve' scheme of the world. This point will be elaborated later in the present paper. What I wanted to do at the present moment was to prevent a premature condemnation of the Presocratics on 'empirical' grounds.

(7) The theory of Thales shares important features both with certain religious doctrines, myths, and with abstract scientific theories. A myth such as the various mythological accounts of the origin of the universe and of the nature of its parts is explanatory. Being concerned with the universe

as a whole rather than with a particular region of it<,> <it> is also general. It may be counterintuitive to those who were brought up in a different tradition and it may be counterinductive when first introduced: the spiritual agencies postulated by many myths are not directly observable. It is quite interesting to note, however, that a firm believer will always be able to provide empirical arguments in addition to the considerations of faith. As we shall try to show later on, a myth is by no means a dreamt up construction that is superimposed upon the facts without being in any way connected with them. Quite the contrary, a good myth will be able to cite many facts in its favour and it will sometimes be more firmly rooted in 'experience' than are some highly appreciated scientific theories of today.

I think I should dwell a little longer on this feature of a myth. It is very often assumed that a myth is a subjective phenomenon, related to poetry, an expression of the fantasies, the wishes, and the fears of those believing in it. This suggests that the members of a community governed by a myth are capable of seeing the world as it really is but prefer dreaming instead. It suggests that it would be very simple getting realistic reports from them – all we need is to wake them up from their dreams and ask them what they *now* perceive. The reports given by them under *these* circumstances so it is assumed, would coincide with our own reports. Briefly and crudely: primitive people live in the same world in which we live, but they prefer to dream up stories for their own amusement. I submit that this idea cannot at all account for the tremendous power a myth has over those believing in it. Nor can it account for the fact which came as a surprise to some ethnologists, that the so-called primitive people can argue very well in favour of their beliefs and that they can explain even apparently contradictory evidence in terms of these beliefs. A myth, therefore, is not merely a dream or a piece of poetry whose distinctness from a true account of the universe is clearly realized. *It is itself supposed to be such a true account*, it is supposed to be in agreement with the facts.

Now the surprising thing is that it is very easily possible to find factual support for ideas which prima facie seem to be completely untenable. This is partly due to the astonishing plasticity of the human mind. Just try to put yourself into the conditions of the fifteenth and sixteenth centuries. The belief in the existence of demons was then very widespread in certain parts of Europe. It was preached from the pulpits, it was used to account for strange meteorological phenomena, for diseases in cattle, for the shortage of crop<s>, for impotence, lack of affection, misformation at birth; in short, it was used for explaining almost everything unpleasant in the everyday life of human beings. It was therefore not at all divorced from reality. Quite the contrary: a great deal of the life of ordinary people consisted in actions aimed at reducing the influence of these evil powers. This was the general situation. How did it influence the mind? The answer can be given with the help of a very simple example. Assume that one of your best friends meets you in the street and says to you: 'Why is your nose so red?'. Will it not seem to you now that the whole world is staring at you, and will you not hurry home in order to put matters right as quickly as possible? Now if a harmless *and isolated* joke can change appearances in such a drastic manner, how much greater must be the influence of a system of thought that is comprehensive, that is taught in the very youth, that is driven home <on> any possible occasion? Will not the presence of this system make the world look very different from what it looks <like> to the agnostics of today? Will not forbidden wishes be immediately interpreted as the result of the presence of demons? Will this not create a feeling of guilt that by far surpasses what we feel today on similar occasions? Will it not be *felt* (and not only *assumed*) that there is really no distinction between public action that can be seen by all and private thought that is accessible to God and to oneself only? Will this not further entail that the distinction between dream and imagination on the one side and reality on the other will cease to be as distinct as it is for us, and that dreams will be taken to be real occurrences (even we sometimes mix up dreams and reality)? Moreover, it is well known that tension of the mind may give rise to phenomena like hearing voices, feeling that one's thoughts are being made up for one, that one's limbs move under the influence of agencies different from one's own will, etc. Michelet has given a wonderful description of the stages through which a normal human being, a woman, might be led into a situation where conversations with demons would seem to her to be something very real. To Huxley we owe a similarly perceptive description of the conditions which lead to an experience of reality that is very different from our own. Far from being a figment of the imagination that is clearly opposed to what is known to be the real world a myth is therefore a system of thought which is supported by numerous very direct and forceful experiences, by experiences, moreover, which seem to be far more compelling than the highly sophisticated experimental results upon which modern science bases its picture of the world. Viewed from the background of the usual empirical approach this is a surprising result indeed! It is also the first indication that there must be something amiss with the fairly popular idea that the distinction between a myth and a scientific theory lies in the factual basis of the latter. Let us elaborate this difficulty![2]

---

[2] For the psychological abnormalities connected with the idea of witchcraft <see> Karl Jaspers, *Allgemeine Psychopathologie* (Berlin: Springer-Verlag, 1959) (English translation: *General Psychopathology* (Chicago: Henry Regnery, 1963) (Ed.)). See also J. Michelet, *La Sorciere* (1956); Aldous Huxley, *The Devils of Loudun* (London: Chatto and Windus, 1952). H. C. Lea's *Materials Towards a History of Witchcraft*, 3 vols (Philadelphia: University of Pennsylvania Press, 1939) offers an excellent survey over the literature of the fourteenth, fifteenth, and sixteenth centuries.

(8) According to the *law of inertia*, which is one of the most fundamental laws of modern physics, both celestial phenomena and phenomena on the surface of the earth are such that an object that is not influenced by forces will move in a straight line with constant speed. Now, let us approach this law from a naïvely observational point of view, i.e., let us use our eyes, collect observations, and look at what these naïvely collected observations tell us about the law. Proceeding in this purely 'inductive' fashion, what do we discover? We discover, first, that the identity, asserted by the law, between the behaviour of celestial and terrestrial phenomena is not in agreement with observations. One of the most striking facts about celestial phenomena is their *regularity*: stars never behave like clouds which form and dissolve almost at random. Their motion can be predicted with great precision on the assumption that they move in circles with constant angular velocity. True, there are some stars, the planets, which cannot readily be incorporated into this simple scheme. However, already at Plato's time it was found that a slight extension of it, which allows for the superposition of circular movements, was entirely successful. This provides strong confirming evidence for the assumption that the stars obey laws of their own, the basic elements of these laws being circular motions with constant angular velocity.

Now a look at terrestrial phenomena shows at once that completely different laws are valid here. To start with, no terrestrial motion goes on forever. Secondly, circular motion is never observed. What *is* observed is this: objects which are *not* influenced from the outside either rise (fire, air), or fall (water, earth), and the speed with which they approach the ground, or the sky, is proportional to the amount of earth (fire) present in them. Objects which *are* influenced from the outside, such as the piece of chalk which I am now holding between my fingers, follow the 'motor' as long as its influence lasts and then either come to a standstill, or assume one of the 'natural' motions, that is, they either move 'up' or 'down'. Once they have assumed this natural motion their behaviour will depend on the elements of which they are composed. If earth preponderates, then they will fall down, and the more earth they contain the faster they will fall.

This is what naïve observations or naïve 'experiment' tells us, and these results are very different indeed from what is asserted by Galileo's law of inertia. Keep this in mind, please, when somebody wants to tell you that modern science (and the birth of modern science is usually connected with the names of Copernicus, Galileo, Kepler, Newton) started when people stopped speculating and started observing instead. You have just seen what the result of such unthinking observation would be: it would be very different from anything that was held to be true by the founders of modern science. (A little more argument would also show that the immobility of the earth is a natural consequence of the simple and strictly 'observational' theory of knowledge I have just outlined.) What I failed to mention is that the theory of motion I have developed in this purely observational fashion is identical, except for some embroideries which I have intentionally left out, with Aristotle's theory, which was overthrown by Galileo and his followers. Quite clearly this elimination cannot have been based upon observation, or even a 'return to the facts', since the theory to be eliminated was itself very firmly based upon observations. Even worse, our above considerations show that the law of inertia which replaced Aristotle's dynamical scheme is counterinductive, just as was Thales' theory and as are many myths, at least when they are first introduced.

The inductivistic defenders of Newton and Galileo will of course reply that their law is supported by the facts *if we only analyse our observations* and interpret them in a manner that is different from the interpretation adopted by Aristotle and by his followers. For example, the circular motion of the stars must not be ascribed to the stars themselves but must be understood as a result of the rotation of the earth; the movement of the sun along the zodiac must not be ascribed to the sun itself, but to the motion of the earth around it, and so on. This remark is quite acceptable but it makes it now very difficult indeed to distinguish between a scientific theory and a myth, such as the myth of witchcraft<,> whose observational success is the result of just such an interpretation of countless empirical events; for example, illness, effects of sexual frustration, nightmares, schizophrenia, and so forth. We seem to be forced to admit – and this consequence has indeed been drawn by some thinkers – that there is no distinction at all between science and myth except perhaps the date: science is the myth of today, myths were the science of the past. Is this conclusion justified? Has no advance taken place, no intellectual improvement? Is the idea that we have progressed in the direction of greater rationality without any foundation? It is this question we shall examine next.[3]

(9) It is quite correct that scientific theories such as Newton's theory of gravitation and Einstein's theory of general relativity are in many respects very similar to all-embracing mythological schemes. They attempt to deal with everything, they are initially counterinductive, and they precipitate a reinterpretation of observational results which eliminates whatever initial disagreement may exist between the theory and 'the facts'. There are many further similarities such as the inertia of the institutions and of the individuals responsible for teaching the myth (the theory): religious institutions, political institutions try to eliminate opposing views either through force or through persuasion; editors of scientific journals are hesitant to

---

[3] For a more detailed description of the Aristotelian theory <see> Marshall Clagett, *Greek Science in Antiquity* (New York: Collier Books, 1957). <See> also Herbert Butterfield, *The Origins of Modern Science, 1300–1800* (London: G. Bell, 1949).

publish papers which would attract the disapproval of the scientific community. Scientific textbooks explain the theories which are commonly accepted and only rarely mention existing alternatives or weaknesses of the current theories. In a book to be published in the near future my colleague, Professor T. S. Kuhn of the University of California, has discussed many examples which exhibit the conservatism inherent in the scientific enterprise and the many similarities which exist between the community of scientists and religious communities. However, despite all these very surprising similarities whose existence flatly refutes most of the current empirical accounts of science, there are features which separate the two fields, or at least should separate them, if the claim for rationality and open-mindedness is justly applied to the sciences.

As will be shown presently, the distinction lies in the psychological attitude of the people who have accepted a myth or a scientific theory as well as in certain peculiarities of the logical structure of both. In the case of Thales and of his successors this distinction was crystal clear. By and large it is retained by the sciences, although it must be admitted that it is not always drawn in a clear fashion and hence is difficult to recognize. We now turn to a more detailed discussion.

(10) The psychological attitude of those believing in a myth can be characterized as an *attitude of complete and unhesitating acceptance*. The myth relates the truth and cannot possibly be mistaken. If trouble arises in the attempt to apply it to reality or to understand reality in its terms then this shows a weakness not of the myth itself but rather of the human beings who use it, who fail to 'see' the clear message it conveys or who fail to conform to the demands the myth imposes upon them. The myth itself is infallible. This attitude is very often supported by certain ideas concerning the origin of the myth: it has been given to mankind by the gods; or by wise men of a time long gone by when the truth was still unconcealed. The manner of teaching, too, is profoundly influenced by this attitude: the teacher is the man who understands and who is therefore completely superior; the pupil is devoid of knowledge (or, at least, of manifest knowledge). He is incapable of judging anything the teacher says. He must therefore remain completely passive. (This raises, of course, the problem how he can possibly grasp anything that is presented to him. This problem cannot be answered by the doctrinaire. But we shall not dwell here on this difficulty.)

The method of teaching corresponding to this idea of the teacher-pupil relationship is a method of indoctrination. At any stage of the indoctrination the progress of the pupil may be tested with the help of questions, or of tasks whose procedure is laid down in advance. The splendour of rituals is frequently used in order to create submissiveness. As has been discovered by psychologists and described in a most interesting manner by Sargant, such rituals are extremely well suited for irreversibly implanting into the mind some important parts of the accepted myth. They use shock tactics and increase fear and despair to an extent that life seems hardly possible, and they introduce the doctrine after the moment of shock, when complete exhaustion has taken over and when the desire for help and understanding has become extreme. In this way a very strong psychological connection is established between one's personal needs and the content of the doctrine. Crafty preachers have always used the fear of hell for the very same purpose in their sermons. I really cannot help being surprised at the large extent to which all these features are preserved in our modern civilization and at places which prima facie would seem to be sources of progress and reason. Is it not true that many university courses proceed in an utterly dogmatic fashion? Is it not true that examinations are designed to test not the pupil's capability for independent thought but the extent to which he has mastered a doctrine, such mastery being regarded as the only thing of importance? Is it not true that even physics textbooks are sometimes written in a fashion which suggests that we have here the final truth which may perhaps by modified in some technical details but whose basic principles are beyond doubt? Is it not true that at least <in> some places radically different accounts are viewed with suspicion or, even worse, as an unwanted and unwarranted disturbance of the peace that follows the discovery of the truth? Are not many events at the universities designed to intimidate and to increase the desire for acquiring what is being offered here? Our contemporary society has indeed many connections still with its primitive ancestors.

The role of moral ideas in the context of a myth is of greatest interest for it is here that support is established in the most subtle and the most misleading fashion. There are obvious cases where the commandments of the ethical code are directly connected with the preservation of the existent order: instruction concerning the behaviour towards the king or the tribal elders, marriage taboos, punishments inflicted for violating the social order are of this kind. It should be observed, by the way, that these cases are distinct from contemporary social morals not only by their content. It is demanded that the king be revered. A violation of this commandment is not merely wicked, it is unnatural, it is a violation of the order of nature which must lead to punishment *as the result of a natural process* rather than of a *decision* made by a group or by an individual. Those who act in order to preserve the natural order are doing so not because they have decided in favour of a closed society and have agreed on the measures necessary to bring it about (such decisions can occur only later when it is attempted consciously to return to a closed society). Their actions appear to them to be the inevitable result of the breach of the ethical code; they are therefore not responsible for it. The domain of human responsibility and the domain of human freedom corresponding to it is consequently much restricted. It is as if a large part of human consciousness, or of the human mind, had turned into a mechanism that works automatically for the preservation of the society. This mechanism is harder than the hardest kind of matter, being supported, as it is, by human belief and, as we shall show later, by arguments leading to absolute certainty. No material is more rigid than the ideas constituting a system of dogmatic belief.

This apologetic function of ethical norms may be present even in those cases where the connection between the myth and the supporting ethics is not immediately obvious. A moral code which demands that due respect be paid to the tribal king and which threatens those who disobey with destruction is quite clearly apologetic. However, even an ethics of virtue, which prima facie seems to be directly connected with the welfare and the improvement of human beings and which refuses to bow to exterior considerations, may be nothing but a cleverly concealed attempt to enlist the most noble feelings for the preservation of some mythological point of view. A good indication of such dishonest use of humanitarian sentiments is the existence of so-called 'higher values'. An ethical system may begin by pointing out that proper attention must be paid to human beings. For example, it may begin by demanding that people should take care of their souls. This is a noble sentiment indeed – but all depends now how the demand is further developed. Assume it is connected with a theological doctrine of the soul. 'Take care of your soul' now means 'obey the laws of the lord'. Which laws? Those laws which are taught by a certain religion. How are these laws to be understood? They are to be understood in the ways the Church Fathers understood them. (Roman Catholic version). Who knows the opinion of the Church Fathers? The present representatives of the Church. Questions of this kind may reveal that the demand 'take care of your souls' means nothing but 'support the church and accept her teachings'. This is how an apologetic ethics may be dressed up as being based upon humanitarian principles.

It is now easy to see how 'higher values' come into existence. A person who interprets the demand 'take care of your souls' in too concrete and immediate a fashion, for example, who interprets it as meaning that people should live under the proper material conditions such that their kids can be free from worry and fatigue, will of course not support the myth. He may be of great service for mankind, he may fight ignorance, poverty, unhappiness; he will still be of no use for strengthening the myth. By directing people's attention away from the myth he may be even potentially dangerous. The best way to eliminate such a person is to discredit him by discrediting what he is doing; he wants to eliminate poverty and hunger – this means he is only interested in the lower and sensual parts of a human being, he is a crude materialist. He wants to eliminate prejudice – this means he must think very highly of his own intellectual abilities, he is conceited. He fights for freedom and for the self-sufficiency of the individual, that is, he is an egoist. Only he who really cares for the soul, who has freed himself from the senses, who has overcome his conceit and his egoism, only he therefore who supports the myth is worthy of being called a good man. The egoist on the other hand is dangerous. He must be severely dealt with, for he corrupts the soul. In this way some of the greatest benefactors of mankind have been accused of being its greatest enemies and have been murdered in the name of the very same humanitarian feelings that guided them in their own pursuits. There is no sadder spectacle than the way in which the most noble sentiments have been misused for protecting and enforcing the most absurd and inhuman doctrines.[4]

(11) The picture which we have drawn so far is this: a myth is a system of thought, possibly false, or perhaps very unsatisfactory from an intellectual point of view, that is imposed and preserved by indoctrination, fear, prejudice, deceit. The idea might now arise that the removal of the psychological forces working in favour of the myth must at once lead to the realization that one has been the victim of a chimera. Nothing could be further from the truth. This has already been pointed out in section 7. The power of a myth is not at all exhausted by the psychological factors we have described so far. A myth can very well stand on its own feet. *It can give explanations*, it can reply to criticism, it can give a satisfactory account even of events which prima facie seem to refute it. It can do this *because it is absolutely true*. It has therefore, something to offer. It has to offer truth, absolute truth. It is for this reason, so the defender of a myth will point out, that all the measures for its preservation have been taken. The preservation of the truth is after all a worthy cause. This is a very powerful argument. It is this argument that has been used again and again to justify the most inhuman measures. We do not really understand a myth unless we accept its claim for absoluteness and try to find out how it is realized.

(12) Such an examination will show, among other things, that the systems of thought of even some very primitive societies are the result of an astounding ingenuity. Far from being incoherent collections of crazy ideas which are only slightly connected with reality, they are logical structures of great sophistication which achieve a very surprising thing: they remain intact in <face> of almost every difficulty. And this is done not just by ignoring the difficulty but in a much more interesting fashion; it is done by turning the difficulty to their own advantage. In order to show the logic of the situation most clearly let us consider the assertion, A, that a person P

---

[4] <See> W. Sargant, *Battle for the Mind: A Physiology of Conversion and Brain-Washing* (London: Heinemann, 1957); E. H. Lecky, *History of the Rise and Influence of the Spirit of Rationalism in Europe, volume 1* (London: Longmans, Green, 1865).

hates his father. Now as long as P is not too friendly to his father A will not be in trouble. However, assume that P is very kind to his father, buys him expensive presents, is most concerned about his welfare. Let O be the description of this behaviour. Does O refute A? It need not if the additional assumption is made, B, that people who hate when they are expected to love feel guilty and try to alleviate their guilt feelings by overcompensation. For now a defender of A may reason as follows: 'He buys presents for his father, he always hangs around him. How very strange. Why should he do such a thing? Does this now show that something fishy is going on? And, of course, I know very well what is going on! He is protesting too much – he overcompensates. But why should he overcompensate unless he feels guilty? And why should he feel guilty unless there is some reason for it – and this reason is that he hates his father, something I have suspected all along.'

Look at the structure of this argument. A is proposed. O turns up and threatens to refute A. Now B is proposed. B gives a new account of O according to which O now supports rather than eliminates A.

Now consider a pair of sentences, A and B, or the simple 'theory' consisting of A and B. Is this theory vulnerable <to> experience? All depends on how the hypothesis B is further explained. It may assert that overcompensation has certain consequences, α, β, γ, ... which are different from A. In this case A & B entails α & β & γ ... and *may* be refuted by $\neg a$ (provided one does not attempt to rescue B by proposing another hypothesis, C, which turns 7a into confirming evidence for B). The 'theory' A & B is then not absolutely certain, it may run into trouble. But assume that A, the assumption reached by B, is regarded as B's only consequence (apart from O). This need not be immediately evident. It will not be if B is connected with other statements which then in various steps lead back to A. However that may be, the situation is now as follows: if P behaves badly towards his father, then A will obviously be supported. If he behaves well, then A will also be supported because of B. The existence of B leads to further predictions, hence A & B is safe *whatever happens* (whether he behaves well or does not behave well), i.e., A & B is absolutely true.

It is easily seen that this absolute truth is due to the way in which A and B are related to each other. Whenever A gets into trouble B comes to the rescue. B on the other hand cannot get into trouble as it is connected with A only and this statement cannot get into trouble. But it is also clear *that we are not forced to relate A and B to each other in that fashion*. As has been said above B may be formulated in a manner which leads to predictions other than those made by A. In this case A & B may well be refuted by further investigation.

What we have arrived at now is an extremely important result. It shows that the certainty possessed by a theory is entirely man made. It is due to the way in which the parts of a theory have been related to each other and to the meanings given to these parts. We shall return to this result later when discussing alternatives to the mythological 'way of life' or the dogmatic way of life, as we shall also call it. What we must realize here is that the claim that a myth offers absolute truth is indeed correct. Of course, a myth will usually be much more complex than the simple theory we have just described. It will contain many elements but these elements will again be related to each other in such a manner that the result is the preservation, and even confirmation, of the myth under all possible circumstances.

The use of closed systems of explanation of the kind just described is not at all restricted to 'primitive' societies. Quite the contrary, there exist very influential theories *today* which work according to very similar principles. Examples are certain parts of the Roman Catholic doctrine (to mention only one of the numerous religious doctrines in existence), certain political theories (here Marxism seems to be a good example), psychoanalysis as well as certain parts of the quantum theory of today. It has been realized by some investigators that psychoanalysis and related theories of the mind are built in a fashion that enables them to take care of almost every difficulty that might arise. Thus Freud's theory of dreams with its distinction between the latent dream and the manifest dream content and the function it ascribes to the dream-censor is very often used for averting a criticism concerning certain assumptions about the way in which people think. I do not mean by this that Freud's theory of dreams is itself dogmatic, or bound to be dogmatic; quite the contrary, I think it contains many interesting observations and is certainly worth being studied most seriously. What I am discussing here is not the *truth* of this theory but rather the way in which it is made to collaborate with other ideas, creating thus a comprehensive theory of the mind which *in toto* is safe from refutation. That is, what I am discussing here is the way in which assumptions which are *possibly true* (but also possibly false) are interlocked inside a more comprehensive theory which, because of this particular form of collaboration and mutual support, will be absolutely true.[5]

(13) To sum up. Myths are regarded with an attitude of complete and unhesitating acceptance. This attitude is reinforced both materially and spiritually. Materially it is reinforced by institutions which we may now summarily describe as totalitarian. Even the arts, mainly the dramatic arts and the way in which they are performed, play an important role here: by creating in the public the impression of necessity, a drama may paralyse the imagination and restrict thought and emotion to a single track or a single pattern. On the spiritual level a myth is reinforced by an ethical code

---

[5] For a description, from the point of view of the text, of some very interesting myths cf. <see> E. Evans-Pritchard, *Witchcraft, Oracles and Magic Among the Azande* (Oxford: Clarendon Press, 1937).

which rewards conformity and punishes opposition. The values of such a code will therefore be very different from the values of a simple humanitarian ethics. Such values are often advertised as 'higher values' which are much more noble than is the materialistic concern for human happiness and the basely rationalistic concern for freedom and independence of thought. A myth is also capable of proving its validity on a purely intellectual and argumentative level. Being built in such a manner that any difficulty can be taken care of and turned into further proof of its excellence it can rightly assert its own absolute truth. Add<ing> to this the change of human life brought about by the attempt to live in accordance with the prescriptions of the myth, the effects of the beliefs held on perception, we see that a myth is the most solid and immutable entity that can be imagined. Each attack increases its solidity by giving rise to explanations which reveal that the attack actually *confirms* the myth. It cannot be destroyed by nature as it turns every feature of nature into proof of its validity. It cannot be destroyed by man either, for the fate of those few who can free themselves from its fetters can easily be foreseen. Quite apart from the fact that society will take measures either to 'educate' or to eliminate them, they will most of the time lack the strength to go against the received opinion which, because of its complicated character and because of its success and the explanations it provides, will seem to embody reason itself. A myth also changes those believing in it to such an extent that they cannot even imagine a world that looks different – if they dare to try it, that is. And yet despite this tremendous power of institutional pressure, fear, spiritual corruption, and ingrained habits of thinking, and despite the splendour of a system that can give security by giving absolute truth, there have always been men who looked for something less impressive, for something 'lower' (as measured by the code of the ad hoc ethics), for something more human. The Ionians did not even have to look – for them this more human form of life was a matter of course. I shall now describe the way in which their theories and their general outlook on life differs from what has just been described.

(14) The Ionians invented cosmological theories and made suggestions for the improvement of the society in which they lived. For them the human origin of explanatory systems and of social laws was, therefore, quite obvious – after all, they had created many of these laws and these explanations themselves, opposing them consciously to the incomplete and unsatisfactory previous accounts. Nor was there any inclination to attribute inventions to distant predecessors, or to the gods – the general attitude was too individualistic for that. It is true that later on many achievements were ascribed to one outstanding individual, to Thales, to Solon and, above all, to Homer. But these individuals are not represented as supermen. They are represented as possessing human faults as is indicated by the many

anecdotes that were current about them. Thus 'a witty and attractive Thracian servant girl is said to have mocked Thales for falling into a well while he was observing the stars and gazing upwards; declaring that he was eager to know the things in the sky, but that what was behind him and just by his feet escaped his notice' (Plato's report). There is no reason to suppose that the story is true – still it indicates the attitude adopted towards these early thinkers: they are not regarded with breathless awe. They are perhaps appreciated, but this admiration does not get out of proportion but is balanced by a good dose of mockery.

This, of course, generates a very different point of view concerning knowledge and society. Theories, the laws of society, are man made. They may have been invented by excellent thinkers – still they participate in the fallibility of all human beings. They may contain errors. It is very likely that they do. They may be in need of improvement. Hence, whatever difficulty is observed in the course of the application of the current knowledge is regarded as a discovery, not of a lack of understanding on the part of those using this knowledge, but rather as an indication of the inner weakness of this knowledge itself. Moreover, if the current cosmology is something that may be in need of improvement then this improvement is better carried out as soon as possible. It can be carried out only if its shortcomings are known. How can they be discovered? By a process of rational criticism which relentlessly investigates every aspect of the theory and changes it in case it is found to be unsatisfactory. The attitude towards a generally accepted point of view such as a cosmological theory or a social system will, therefore, be an *attitude of criticism*. Far from being regarded as the final truth in all matters such a point of view will be seen as a first step whose limitations must be found in order that improvement become<s> possible. This is the most fundamental difference between the closed society that is governed by a myth and an open society. Both have created their own social institutions and both have invented explanatory systems in order to account for the features of the world around them. However only in the latter is there an awareness of the human origin of these institutions and these systems and only they are prepared openly to admit it and to draw the necessary consequences: the need for improvement, the need for criticism, the inappropriateness of dogmatism. For a closed society the institutions, the myth, the ethical code will be a datum of a perhaps very elevated origin and this datum must be unconditionally accepted, not a single doubt must be directed against it. Let us now investigate the consequences of this difference of attitude.

The feature which to my mind is the most important one will be the absence of an ethics of apology and of 'higher values'. If the cosmology is of human origin, if the society is of human origin, then the respect paid to it cannot be greater and should not be greater than the respect paid to the human beings who have contributed to its invention. More especially there is no need and no sense in making adherence to the cosmology a virtue that outdoes the simple humanitarian virtue of respect for the individual. This means, of course, that systems of thought and institutions are judged by human standards *and not the other way around.* The excellence of an individual, or his happiness, is not regarded as something altogether inferior to the virtues derived from the acceptance of a cosmological system. It may happen that outstanding individuals are appreciated because of their devotion to spiritual matters, and intelligence may be regarded as a value that is higher than mere bodily excellence. However such judgements now have nothing to do with cosmological systems; they are not made in order to support a certain theory nor do they receive support from such a theory. They stand on their own feet. The importance of this development which may be described as the discovery of the distinction between nature and convention can hardly be exaggerated. It means the development of a self-sufficient ethics; and this in turn means a tremendous extension of the domain of human responsibility. So far man was only responsible for acting or not acting in accordance with a certain code. The code itself was 'given' by the basic myth and is unalterable. Now it is admitted that man is the creator even of the most basic rules of behaviour and therefore responsible for them, too. This is the step from childhood into maturity.

(It ought to be pointed out, by the way, that the Greeks had predecessors in this, even in Egypt. However, there is no time now to discuss these developments in greater detail.)

Ethics was thus freed from cosmology. And the cosmology itself was altered most radically. Ideas are to be changed, investigated, improved, abandoned. This can be done only if their limitations are realized from the very beginning. The teacher therefore ceases to be the oracle he was in the closed society. What he wants to impart <to> the pupil is not a doctrine, but an attitude. He will not hesitate to explain the past and current theories as well as his own opinion. But at the same time he will point out that these theories are only a starting point, that they possess weaknesses and must be improved. He will emphasize that even a theory that now seems to be without blemish is bound to have shortcomings. He will impress upon the pupil that the task is to find out these shortcomings by a process of critical examination. It will soon emerge that such an attitude of criticism is not something purely negative, an attitude of barren scepticism, but that it requires a good deal of imagination and that its application to a particular case can be regarded as successful only if it leads to improvement, to a progress of knowledge. It is very important to realize this side of the critical attitude. It does not consist in the general assertion which we find among some philosophical sceptics that knowledge is impossible. Nor should a proper instruction in this attitude create cynics. Strong beliefs are admitted and even encouraged; enthusiasm for a particular theory is most welcome and so is tenacity in the face of difficulties (after all, the theory *might* be capable of solving them). All that is required is the realization that however basic and obvious a certain belief may seem and however great its appeal to those who have adopted it, *one* of the possible ways of getting out of trouble is the elimination of this belief and its replacement by something very different. Thus it is quite in order for a teacher to explain to a pupil his own convictions and to present them in the strongest and most convincing form. But at the same time it must be very clear that excellence does not mean absolute truth and that even the most beautiful ideas might fail.

Again I cannot refrain from remarking that university courses of today very often proceed in a very different fashion. I do not want to dwell on the so-called humanities where the craving for admiration very often makes agreement with the teacher, total agreement that is, a condition of academic success. One is relieved to observe that many such teachers are not really taken seriously, that the students are very aware what is going on and agree only in order to succeed. (Unfortunately this is not always so as is demonstrated by the existence of schools which centre around the most ridiculous doctrines.) What is really astonishing, however, is the existence of such phenomena in the natural sciences, especially in physics. As has been pointed out by Professor T. S. Kuhn, you do not hear much of the need for criticism in a textbook for physics. Theories are introduced in a fashion which suggests that this is how things are and their weaknesses are rarely mentioned. The straightforwardness of the Ionians in these matters is not often met today (the field theories form perhaps an exception – but this exception occurs on a purely formal level).

So far I have described the attitude towards theories found among the Ionians and the way in which they treated matters of cosmology and matters of ethics. Now in order for a critical attitude to make sense, the theories to which this attitude is applied *must be fallible*, that is, they must be accessible to criticism and they must not be built in a manner which in advance guarantees their absolute validity. The theories connected with the critical attitude must therefore be very different indeed from dogmas and mythological schemes. Their elements must not be interlocked in a fashion which allows the total theory to take care of any imaginable situation; it must not be possible to reinterpret evidence that refutes one postulate as evidence that confirms another and forms the total content of that other postulate. No such collaboration of postulates must occur.

As has been explained above it is possible to create theories which have this property. The simplest way of doing this consists in suggesting independent tests for each assumption that enters a more comprehensive theoretical system. We are therefore able fully to realize all the conditions necessary for the critical way of life. This leads at once to the following fundamental problem: which attitude shall we adopt and which kind of life shall we lead? This is the most fundamental problem of all epistemology.

As is well known, traditional epistemology sets itself the task of finding what it calls the foundations of all our knowledge. By a foundation is meant a more restricted body of theory or factual description which is absolutely certain and such that the total knowledge can be obtained from it in a fairly simple and straightforward fashion. Sense-data, the intuition of clear and distinct ideas, are foundations of knowledge in the sense just explained. Such foundations have exactly the characteristics of a myth as it has been explained above: they are regarded with an attitude of complete acceptance (compare, for example, the almost religious overtones in the empiricist invitation to pay due reverence to the facts); they are received passively; the knowledge derived from them is arranged in a manner which guarantees absolute certainty. This being the case any decision against methods creating certainty will at the same time be a decision against the acceptance of foundations of knowledge in the sense just described; *it will be a decision in favour of a form of knowledge that possesses no foundation.* And it will therefore also be a decision to leave the traditional path of epistemology and to build up knowledge in an entirely new fashion.[6]

(15) It must again be repeated that we are here confronted with a real *decision*, that is, a real choice with a situation which has to be resolved on the basis of our demands and preferences and which cannot be resolved by proof. It is easy to see that these demands and these preferences have quite a lot to do with the welfare of human beings and are therefore ethical demands: epistemology, or the structure of the knowledge we accept, is grounded upon an ethical decision. This result is very different indeed from what seems to be the commonly accepted point of view. For it is usually assumed that the foundations of our knowledge are things which exist independently of human beings, which can be forgotten, misunderstood, overlooked but not eliminated with the help of a decision. This is quite correct *provided* we have already accepted a dogmatic point of view that works with certainties. Such a point of view will, of course, treat its own foundations as something that are given and cannot be influenced by human decisions. However, the dogmatic point of view itself is not given (except, perhaps, historically), it is the result of (conscious or unconscious) measures, institutional, logical and otherwise, and it can be eliminated by taking different measures. Hence we are again called upon to decide what we like better, a theory that is accepted with complete faith, that is

---

[6] <See> chapter 5 of Popper's *The Open Society and Its Enemies, volume 1: Plato* (London: Routledge and Kegan Paul, 1945). This chapter contains an excellent account of the history of the separation of ethical norms and factual descriptions.

constructed in a manner that makes refutation impossible, that infiltrates ethical considerations, or a theory that is regarded critically, that is capable of improvement and which still leaves us the freedom to arrange our lives in the manner we find most congenial.

It has already become clear that a solution of the problem is bound profoundly to influence the institutions of the society. A dogmatic point of view, a myth, will need totalitarian institutions or at least institutions whose main function is to preserve the central doctrine and perhaps to defend it against those who think that a different approach is needed; such a point of view will use indoctrination and will thereby severely curtail the imagination of the young. This imagination is really astounding. But instead of using it for the purpose of advancing our knowledge the dogmatist restricts it and almost completely eliminates it: we learn 'the truth' and have either to forget the rest or to sever it completely from the domain of truth and to hand it over to poetry. This leads to a dry and boring science and an unintelligent poetry. We see already here that the dogmatic approach will not influence the sciences only. It will mould every activity of human life, emotions and thought alike. There will be an ethics of 'higher values' and everything connected with such values such as conceit, cruelty, the feeling of superiority of the 'just', restriction of freedom. And there will be theories which are absolutely true. A person who values human freedom and responsibility, who sees a human being as one and not divided into sections belonging to the feelings <and> to thought respectively, who values imaginativeness and wants to see it developed, will find the critical attitude much more acceptable – and I for one think that these simple considerations must indeed force everybody except those who have already become too rigid to think independently and to rethink their own lives to accept the critical mode of life and all that goes with it.

I do not want to make it appear here that the decision in favour of the critical way of life is a matter of course and that all those who decided against it were quite superficially mistaken. After all this decision involves a complete change of attitude in all fields where certainty is in the centre – in religion, and in traditional philosophy. It will perhaps force us to abandon, or at least to regard with caution, the instruments bringing about this certainty, such as the classical tragedy (but we may retain the classical comedy). Wagner will have to disappear (which at least for this writer is not too tragic an event). Above all we shall have to give up the desire for certainty, the wish to escape our responsibilities. The first one should be easy as soon as it has been realized that certainty is not something that exists independently of the human will but rather a reflection of the way in which we (consciously or unconsciously) proceed. If certainty is our own product then it cannot be worth more than the human beings who created it in the first place. Moreover, there is a very simple logical point which is another argument against the use of certainties in knowledge. It has always been believed that only a point of view that has been established with certainty can be regarded as giving a true account of the world. This is mistaken. Just assume that the world suddenly changes from today to tomorrow. Being built in such a manner that it can take care of everything, the accepted doctrine will not be influenced by this. But this means that it cannot distinguish between the world as it is and the world as it could be. It cannot distinguish between the one real world we live in and the many possible worlds one can dream up, and is therefore completely useless for describing, and thereby singling out the former. It has no relation to anything really existing, it is rather like a dream to which we cling despite the fact that the real world may be very different. Of course, a theory that is fallible is also our product. But *its actual failure* is not any longer our product. This may happen at any time, in the most unforeseen circumstances, and it is this failure which, because of its unforeseen character, brings us in contact with what is independent of us, viz., the real world in which we live. These, I think, are strong arguments in favour of accepting the critical mode of life and thereby renouncing almost all traditional philosophy except that of the Presocratics.

(16) The development of the doctrine of the Presocratics gives an excellent example of how life is and how thinking is in a critical community. Hardly anywhere else in the history of thought do we have such an outstanding variety of theories. Hardly anywhere else is such swift progress made in our knowledge. We start with the idea that everything is basically one kind of substance – water. In less than two hundred years we arrive at the atomic theory which forms still the basis of contemporary physics. The development is by criticism and improvement. Anaximander's theory may be regarded as the result of a criticism of the theory of Thales: the four elements, water, fire, earth and air occur equally frequently in the universe. Why, then, give water the preference? Water is necessary for life. But so is fire, air, earth. All these substances are on the same level. None of them can therefore justifiably be regarded as the one basic substance which underlies all the others. Hence, if there is a basic substance – an assumption which is retained by Anaximander – it cannot be any one of the known elements. Not having been thought of before, it cannot even have a name. Being the common origin of everything determined, it cannot have determined properties either, or at least its properties cannot be those with which we are acquainted from common observation. Anaximander therefore calls his basic substance the *indefinite*. There is no time here to discuss the very interesting ideas Anaximander developed about the manner in which the elements arise out of the indefinite. These ideas have been taken up by Heisenberg and form an integral part of his most recent speculations. Nor can we here draw attention to the very interesting cosmology and

cosmogony which Anaximander has developed. Let me emphasize only one point, the way in which the theory of the indefinite was developed. It was developed by a conscious criticism of the theory of Thales which retained the valuable, or apparently valuable, elements of that theory (assumption of a *single* substance) and changed what was found to be objectionable (identification of this single substance with water). Exactly the same kind of procedure led to the atomic theory. Here we must be even more brief. We cannot even present an outline of the ideas of Parmenides, that most influential of all Presocratic thinkers. All we can do is to discuss the way in which his ideas were interpreted by thinkers like Leukippos (who is reported to have been his pupil) and Aristotle.

According to Parmenides monism and change are incompatible. Change is transition from a certain kind of substance into a different kind of substance. Monism asserts that there is only one kind of substance. Consequence: a monist must either deny the existence of change, or admit that his own theory is incorrect. Parmenides took the first line. We shall not deal with his own theory. The atomists and Aristotle took the second line. They admitted the incorrectness of monism and developed pluralistic theories. Aristotle's theory had the advantage of not only making change *possible* (through the pluralism of potentiality and actuality) but of even developing ideas concerning the *causes* of all possible change. It was this theory that was retained through the Middle Ages and whose criticism by the Paris school and the Merton school led finally to modern mechanics. The theory of the atomists was not developed in such detail; it explained the possibility of change (which monism had denied) but did not possess a coherent theory of causes and inertial motions.

It ought to be pointed out, by the way, that atomism and Aristotelianism were not the only answers suggested to the Parmenidean problem. Empedocles, Anaxagoras developed other ideas, there is the theory of Heraklitos and many more. Never before were there so many ideas developed in such a short period of time. But the end was near. For the defenders of the dogmatic point of view were about to commence their counter-attack. This is the beginning of the traditional epistemology with its search for sources of knowledge. There is no time in these two lectures to give even the briefest account of this development. But I shall do this: I shall discuss with you some of the objections which were raised by some Greeks of the Fourth Century against the Ionian philosophers of nature and their achievements. These objections are very clear, very brief. Nothing new has been added to them in the later history of thought. They possess a great intuitive appeal and they are even used in order to lead the general public into some dogmatic belief. They are the most forceful arguments the dogmatic tradition possesses. I shall close my lecture with an account of these arguments and with their refutation.

(17) I begin with two very popular arguments. These arguments occur not only in philosophical writings. We find them also in the Attic comedy which indicates that they must have been known to a great many people. They were used later by the Church Fathers against the ancient philosophers and they are used even today, both by religious groups and by certain schools in physics. The first argument is that the ideas of the early Ionians were *absurd*. Is it not absurd to assume that this piece of wood which I hold in my hand is made of water? Is it not absurd to assume that everything is always in motion, the flame of the candle as well as the steel pen with which I am writing? Those of you who have some acquaintance with the most 'modern' linguistic philosophy will recognize that this argument puts in the 'material mode of speech' (i.e., by talking about things and their properties) what linguistic analysis and especially that brand of linguistic analysis that makes the 'ordinary language' its measure of reason (I do not pretend to know what this 'ordinary language' is) repeats in the 'formal mode of speech' (i.e. by talking about words and their relations). And ordinary language philosophers have indeed repeated the argument from absurdity by pointing out that making my steel pen consist of water involves a misuse of the word 'water'.

It is quite obvious that this argument cannot create the slightest difficulty. What does it mean to say that a certain theory is absurd? It means that the theory runs counter to some very popular ideas with which we are so familiar that we regard them as the expression of the obvious. But this is exactly as it ought to be. After all, these early thinkers wanted to make progress; they wanted to improve our knowledge. Improvement may mean change, very radical change. The charge of absurdity indicates that a very radical change has indeed taken place. Was it a change for the worse? The argument assumes that it was, taking it for granted that in matters of common interest common sense has already arrived at the truth. Such presumption may be forgiven the unreflecting popular critics of the Ionians. However, it is very surprising to find the very same assumption accepted by philosophers who use any opportunity to proclaim their own modernity: all the subtle and boring analyses of the various linguistic schools of today, Wittgensteinians included, proceed from the assumption that the common idiom and the common belief behind it is a good basis for philosophy. It does not need a philosopher to realize how mistaken this idea is.

The second argument proceeds from the astounding variety of ideas produced by the early Ionians. It interprets this variety as an indication that the truth has not yet been found and that it cannot be found in the fashion of the Ionians. What are these philosophers doing? Each of them develops his own pet ideas without regard for the truth. This argument occurs again and again in the history of philosophy. Plato has made full use of it. The Church Fathers use it in order to show the confusion of all philosophy. The empiricists of the seventeenth and early eighteenth centuries use the very same argument in order to show where one gets when leaving, this time not the Bible, but the solid ground of experience. And some contemporary physicists compare the monolithic unity of the Copenhagen point of view with the variety of ideas discussed in the opposite camp. In all these cases uniformity is regarded as a virtue, and variety of ideas, wealth of ideas, imagination as a vice. Such a manner of thinking quite obviously assumes that the truth is best caught with the help of an unchanging single theory, that is, with the help of a myth. In the last section we have given arguments why a myth should not be used for the purpose of explanation. This disposes of the second argument.

An additional observation is in order. In the second argument the theories of the early Ionians are represented as being completely unrelated to each other. This is not correct. As has been pointed out above there is a definite line of development: later theories are the result of a criticism of the early theories and of their improvement in the light of this criticism. Atomism cannot be imagined without Parmenides. Thus far from finding the chaos the critics assert exists we find order and progress through reason.

The two arguments we have discussed so far have been fairly popular. They have been used by philosophers and by laymen alike. The arguments to follow are more esoteric, but equally simple to refute. Assume somebody points out that certainty is an essential part of knowledge in the sense that the meaning of the word 'knowledge' contains the idea of certainty. The answer is very simple: we have decided against certainty (see the arguments in the last section). We have thereby also decided against knowledge in the sense alluded to. If certainty is part of knowledge, then we simply do not want to know in this sense. It has to be admitted that the word 'knowledge' has a very attractive sound to it. But in philosophical matters the sound of a word has no force whatever.

Another criticism consists in pointing out that these early thinkers were very naïve. They undertook the task of constructing theories about the universe as a whole without first having investigated whether the human mind is at all capable of possessing knowledge of such a kind. The reply to this criticism is very obvious: the history of thought has shown that the human mind *is* capable of developing such theories. Moreover, how are we supposed to decide about the ability, or the lack of ability, of the human mind to recognize the universe? Is it not assumed by the argument that a theory of the mind will be simpler than cosmology? Is this not itself a very questionable assumption? At the bottom of all this there is, of course, again the idea of certainty: we can be certain about the features of our mind to which we have direct access. We cannot be so certain about the stars which, after all, are very far away. Nothing could be further from the truth.

Is it not the case that an astronomer may be better acquainted with some galaxy of his choice than with his own ideas concerning a certain lady? Such situations are known to everybody. To give priority to the mind therefore solves nothing.

Practicality, or the absence of it, is another criticism which was frequently directed against the Ionians and against all theoretical investigations. Prima facie this argument is very forceful: what has the fact that everything consists of water got to do with our happiness? The answer is that we do not know. What we *do* know is that participating in the activity of suggesting and criticizing theories will tremendously develop the human imagination, will set the human mind free. And it is impossible to increase happiness without possessing some imagination.

(18) Time has now come to sum up and to present my main thesis. I have shown that there are at least two possible forms of life and that both of them are connected with certain forms of knowledge. I have shown that the choice between these two forms of life is a genuine choice which must be made individually by everybody who is presented with it on the basis of his own demands and his own ideas. I have also given my reasons in favour of choosing the critical mode of life. I think that these reasons are compelling. I realize, however, that what I use as a basis for making my decision may be utterly repulsive for somebody else and may be a reason for him to choose the dogmatic form of life. I have hinted that choice of the critical form of life will have repercussions in very distant fields, for example, in the dramatic arts. Today the theatre corresponding to the critical form of life is hardly existent, but we may hope that Brecht's *theory* (his *plays* seem to have been a failure in this respect) will give rise to experiments leading in the right direction. And then, of course, there is the Attic comedy as an exquisite example. Poetry, too, may have to undergo decisive changes. All this is still *Zukunftsmusik*, and this because our contemporary society is still in many respects very similar to the tribal societies of the past. Philosophy, and especially epistemology, has been trying for the past 2,000 years to tie down the flight of the imagination by referring to 'sources' of knowledge. A tremendous amount of sophistication has been spent, a tremendous amount of energy has been wasted. The greatest thinkers, men with beautiful and exciting ideas, felt that having ideas was just not enough and that the ideas had to be turned into a myth in order to be really acceptable. The task of epistemology has always been to carry out this transformation with the help of arguments of a sometimes very formal character (I am thinking here of the so-called 'philosophy of science' of today). All this is of no avail. What is needed is a completely new start or, more correctly, a return to the freedom of theorizing that was invented by the Ionian philosophers of nature.

3

How to be a good empiricist: a plea for tolerance in matters epistemological

> 'Facts?' he repeated. 'Take a drop more grog, Mr Franklin, and you'll get over the weakness of believing in facts! Foul play, Sir!'
>
> Wilkie Collins *The Moonstone*

I CONTEMPORARY EMPIRICISM LIABLE TO LEAD TO ESTABLISHMENT OF A DOGMATIC METAPHYSICS

Today empiricism is the professed philosophy of a good many intellectual enterprises. It is the core of the sciences, or so at least we are taught, for it is responsible both for the existence and for the growth of scientific knowledge. It has been adopted by influential schools in aesthetics, ethics and theology. And within philosophy proper the empirical point of view has been elaborated in great detail and with even greater precision. This predilection for empiricism is due to the assumption that only a thoroughly observational procedure can exclude fanciful speculation and empty metaphysics as well as to the hope that an empiristic attitude is most liable to prevent stagnation and to further the progress of knowledge. It is the purpose of the present paper to show that empiricism in the form in which it is practised today cannot fulfil this hope.

Putting it very briefly, it seems to me that the contemporary doctrine of empiricism has encountered difficulties, and has created contradictions which are very similar to the difficulties and contradictions inherent in some versions of the doctrine of democracy. The latter are a well-known phenomenon. That is, it is well known that essentially totalitarian measures are often advertised as being a necessary consequence of democratic principles. Even worse – it not so rarely happens that the totalitarian character of the defended measures is not explicitly stated but covered up by calling them 'democratic', the word 'democratic' now being used in a new, and somewhat misleading, manner. This method of (conscious or unconscious) verbal camouflage works so well that it has deceived some of the staunchest supporters of true democracy. What is not so well known is that modern empiricism is in precisely the same predicament. That is, some of the methods of modern empiricism which are introduced in the spirit of anti-dogmatism and progress are bound to lead to the establishment of a dogmatic metaphysics and to the construction of defence mechanisms which make this metaphysics safe from refutation by experimental inquiry. It is true that in the process of establishing such a metaphysics the words 'empirical' or 'experience' will frequently occur; but their sense will be as distorted as was the sense of 'democratic' when used by some concealed defenders of a new tyranny.[1] This, then, is my charge: far from eliminating dogma and metaphysics and thereby encouraging progress, modern empiricism has found a new way of making dogma and metaphysics respectable, viz., the way of calling them 'well-confirmed theories', and of developing a method of confirmation in which experimental inquiry plays a large though well controlled role. In this respect, modern empiricism is very different indeed from the empiricism of Galileo, Faraday and Einstein, though it will of course try to represent these scientists as following its own paradigm of research, thereby further confusing the issue.[2]

From what has been said above it follows that the fight for tolerance in scientific matters and the fight for scientific progress must still be carried on. What has changed is the denomination of the enemies. They were priests, or 'school-philosophers', a few decades ago. Today they call themselves 'philosophers of science', or 'logical empiricists'.[3] There are also a good many scientists who work in the same direction. I maintain that all these groups work against scientific progress. But whereas the former did so openly and could be easily discerned, the latter proceed under the flag of progressivism and empiricism and thereby deceive a good many of their followers. Hence, although their presence is noticeable enough they may almost be compared to a fifth column, the aim of which must be exposed in order that its detrimental effect be fully appreciated. It is the purpose of this paper to contribute to such an exposure.

---

[1] K. R. Popper, *The Open Society and its Enemies* (London: Routledge and Kegan Paul, 1945, reprinted, New Jersey: Princeton University Press, 1953).

[2] It is very interesting to see how many so-called empiricists, when turning to the past, completely fail to pay attention to some very obvious facts which are incompatible with their empiristic epistemology. Thus Galileo has been represented as a thinker who turned away from the empty speculations of the Aristotelians and who based his own laws upon facts which he had carefully collected beforehand. Nothing could be further from the truth. *The Aristotelians could quote numerous observational results in their favour.* The Copernican idea of the motion of the earth, on the other hand, did not possess independent observational support, at least not in the first 150 years of its existence. Moreover, it was inconsistent with facts and highly confirmed physical theories. And *this* is how modern physics started: not as an observational enterprise *but as an unsupported speculation that was inconsistent with highly-confirmed laws.* For details and further references see my 'Realism and Instrumentalism', to appear in M. Bunge (ed.), *The Critical Approach to Science and Philosophy: Essays in Honor of Karl Popper* (New York: Free Press, <1964>).

[3] One might be inclined to add those who base their pronouncements upon an analysis of what they call 'ordinary language'. I do not think they deserve to be honoured by a criticism. Paraphrasing Galileo, one might say that they 'deserve not even that name, for they do not talk plainly and simply but are content to adore the shadows, philosophizing not with due circumspection but merely from having memorized a few ill-understood principles'.

I shall also try to give a positive methodology for the empirical sciences which no longer encourages dogmatic petrification in the name of experience. Put in a nutshell, the answer which this method gives to the question in the title is: you can be a good empiricist only if you are prepared to work with many alternative theories rather than with a single point of view and 'experience'. This plurality of theories must not be regarded as a preliminary stage of knowledge which will at some time in the future be replaced by the One True Theory. Theoretical pluralism is assumed to be an *essential feature* of all knowledge that claims to be objective. Nor can one rest content with a plurality which is merely abstract and which is created by denying now this and now that component of the dominant point of view. Alternatives must rather be developed in such detail that problems already 'solved' by the accepted theory can again be treated in a new and perhaps also more detailed manner. Such development will of course take time, and it will not be possible, for example, at once to construct alternatives to the present quantum theory which are comparable to its richness and sophistication. Still, it would be very unwise to bring the process to a standstill in the very beginning by the remark that some suggested new ideas are undeveloped, general, metaphysical. *It takes time to build a good theory* (a triviality that seems to have been forgotten by some defenders of the Copenhagen point of view of the quantum theory); and it also takes time to develop an alternative to a good theory. The *function* of such concrete alternatives is, however, this: they provide means of criticizing the accepted theory in a manner which goes *beyond* the criticism provided by a comparison of that theory 'with the facts': however closely a theory seems to reflect the facts, however universal its use, and however necessary its existence seems to be to those speaking the corresponding idiom, its factual adequacy can be asserted only *after* it has been confronted with alternatives *whose invention and detailed development must therefore precede any final assertion of practical success and factual adequacy*. This, then, is the methodological justification of a plurality of *theories*: such a plurality allows for a much sharper criticism of accepted ideas than does the comparison with a domain of 'facts' which are supposed to sit there independently of theoretical considerations. The function of unusual *metaphysical* ideas which are built up in a nondogmatic fashion and which are then developed in sufficient detail to give an (alternative) account even of the most common experimental and observational situations is defined accordingly: they play a decisive role in the criticism and in the development of what is generally believed and 'highly confirmed'; and they have therefore to be present at *any* stage of the development of our knowledge.[4] A science that is free from

---

[4] It is nowadays frequently assumed that '[i]f one considers the history of a special branch of science one gets the impression that non-scientific elements ... relatively frequently occur in

metaphysics is on the best way to becoming a *dogmatic* metaphysical system. So far the summary of the method I shall explain, and defend, in the present paper.

It is clear that this method still retains an essential element of *empiricism*: the decision between alternative theories is based upon *crucial experiments*. At the same time it must *restrict* the range of such experiments. Crucial experiments work well with theories of a low degree of generality whose principles do not touch the principles on which the ontology of the chosen observation language is based. They work well if such theories are compared with respect to a much more general background theory which provides a stable meaning for the observation sentences. However, this background theory, like any other theory, is itself in need of criticism. Criticism must use alternatives. Alternatives will be the more efficient the more radically they differ from the point of view to be investigated. It is bound to happen, then, that the alternatives do not share a single statement with the theories they criticize. Clearly, a crucial experiment is now impossible. It is impossible, not because the experimental device is too complex, or because the calculations leading to the experimental prediction are too difficult; it is impossible because there is no statement capable of expressing what emerges from the observation. This consequence, which severely restricts the domain of empirical discussion, cannot be circumvented by any of the methods which are currently in use and which all try to work with relatively stable observation languages. It indicates that the attempt to make empiricism a universal basis of all our factual knowledge cannot be carried out. The discussion of this situation is beyond the scope of the present paper.

On the whole, the paper is a concise summary of results which I have explained in a more detailed fashion in the following essays: 'Explanation, Reduction, and Empiricism'; 'Problems of Microphysics'; 'Problems of Empiricism'; 'Linguistic Philosophy and the Mind-Body Problem'.[5] All the relevant acknowledgements can be found there. Let me only repeat here that my general outlook derives from the work of K. R. Popper (London) and David Bohm (London) and from my discussions with both. It was severely tested in discussion with my colleague, T. S. Kuhn (Berkeley). It

the earlier stages of development, but that they gradually retrogress in later stages and even tend to disappear in such advanced stages which become ripe for a more or less thorough formalization'. (H. J. Groenewold, 'Non-Scientific Elements in the Development of Science', *Synthese*, vol. 10, 1957, p. 305). Our considerations in the text would seem to show that such a development is very undesirable and can only result in a well-formalized, precisely expressed, and completely petrified metaphysics.

[5] These essays were published in Volume III of the *Minnesota Studies in the Philosophy of Science*; in Volumes I and II of the *Pittsburgh Studies in the Philosophy of Science*; and in *Problems of Philosophy, Essays in Honor of Herbert Feigl*, respectively. [In fact, the last paper referred to here never appeared. Feigl's *festschrift* is Feyerabend and Maxwell (1966). (Ed.)].

was the latter's skilful defence of a scientific conservatism which triggered two papers, including the present one. Criticisms by A. Naess (Oslo), D. Rynin (Berkeley), Roy Edgley (Bristol) and J. W. N. Watkins (London) have been responsible for certain changes I made in the final version.

2 TWO CONDITIONS OF CONTEMPORARY EMPIRICISM

In this section I intend to give an outline of some assumptions of contemporary empiricism which have been widely accepted. It will be shown in the sections to follow that these apparently harmless assumptions which have been explicitly formulated by some logical empiricists, but which also seem to guide the work of a good many physicists, are bound to lead to exactly the results I have outlined above: dogmatic petrification and the establishment, on so-called 'empirical grounds' of a rigid metaphysics.

One of the cornerstones of contemporary empiricism is its *theory of explanation*. This theory is an elaboration of some simple and very plausible ideas first proposed by Popper[6] and it may be introduced as follows: let T and T' be two different scientific theories, T' the theory to be explained or the explanandum, T the explaining theory or the explanans. Explanation (of T') consists in the *derivation* of T' from T and initial conditions which specify the domain D' in which T' is applicable. Prima facie, this demand of derivability seems to be a very natural one to make for 'otherwise the explanans would not constitute adequate grounds for the explanation' (Hempel).[7] It implies two things: first, that the consequences of a satisfactory explanans, T, inside D' must be compatible with the explanandum, T'; and secondly, that the main descriptive terms of these consequences must either coincide, with respect to their meanings, with the main descriptive terms of T', or at least they must be related to them via an empirical hypothesis. The latter result can also be formulated by saying that the meaning of T' must be unaffected by the explanation. 'It is of the utmost

[6] See K. R. Popper, *The Logic of Scientific Discovery* (London: Hutchinson, 1959), section 12. This is a translation of his *Logik der Forschung* published in 1935. The decisive feature of Popper's theory, a feature which was not at all made clear by earlier writers on the subject of explanation, is the emphasis he puts on the initial conditions and the implied possibility of two kinds of laws, viz. (1) laws concerning the temporal sequence of events; and (2) laws concerning the space of initial conditions. In the case of the quantum theory, the laws of the second kind provide very important information about the nature of the elementary particles and it is to *them* and *not* to the laws of motion that reference is made in the discussions concerning the interpretation of the uncertainty relations. In general relativity, the laws formulating the initial conditions concern the structure of the universe at large and only by overlooking them could it be believed that a purely relational account of space would be possible. For the last point, cf. E. L. Hill, 'Quantum Physics and the Relativity Theory', in H. Feigl and G. Maxwell (eds.), *Current Issues in the Philosophy of Science* (New York: Holt, Rinehart and Winston, 1961).
[7] C. G. Hempel, 'Studies in the Logic of Explanation', reprinted in H. Feigl and M. Brodbeck (eds.), *Readings in the Philosophy of Science* (New York: Appleton-Century-Crofts, 1953), p. 321.

importance', writes Professor Nagel,[8] emphasizing this point, 'that the expressions peculiar to a science will possess meanings that are fixed by its *own* procedures and are therefore intelligible in terms of its own rules of usage; whether or not the science has been, or will be [explained in terms of] the other discipline'.

Now if we take it for granted that more general theories are always introduced with the purpose of explaining the existent successful theories, then every new theory will have to satisfy the two conditions just mentioned. Or, to state it in a more explicit manner,]

(1) only such theories are then admissible in a given domain which either *contain* the theories already used in this domain, or which are at least *consistent* with them inside the domain;[9] and

(2) meanings will have to be invariant with respect to scientific progress; that is, all future theories will have to be phrased in such a manner that their use in explanations does not affect what is said by the theories, or factual reports to be explained.

These two conditions I shall call the *consistency condition* and the *condition of meaning invariance*, respectively.

Both conditions are *restrictive* conditions and therefore bound profoundly to influence the growth of knowledge. I shall soon show that the development of actual science very often violates them and that it violates them in exactly those places where one would be inclined to perceive a tremendous progress of knowledge. I shall also show that neither condition can be justified from the point of view of a tolerant empiricism. However, before doing so I would like to mention that both conditions have occasionally entered the domain of the sciences and have been used here in attacks against new developments and even in the process of theory construction itself. Especially today, they play a very important role in the construction as well as in the defence of certain points of view in microphysics.

Taking first an earlier example, we find that in his *Wärmelehre*, Ernst Mach[10] makes the following remark:

Considering that there is, in a purely mechanical system of absolutely elastic atoms no real analogue for the *increase of entropy*, one can hardly suppress the idea that a violation of the second law ... should be possible if such a mechanical system were the *real* basis of thermodynamic processes.

And referring to the fact that the second law is a highly confirmed physical

[8] E. Nagel, 'The Meaning of Reduction in the Natural Sciences', reprinted in A. C. Danto and S. Morgenbesser (eds.), *Philosophy of Science* (New York: World Publishing, 1960), p. 301.

[9] It has been objected to this formulation that theories which are consistent with a given explanandum may still contradict each other. This is quite correct, but it does not invalidate my argument. For as soon as a single theory is regarded as sufficient for explaining all that is known (and represented by the other theories in question), it will have to be consistent with all these other theories.

[10] E. Mach, *Wärmelehre* (Leipzig, 1897), p. 364.

law, he insinuates (in his *Zwei Aufsätze*)[11] that for this reason the mechanical hypothesis must not be taken too seriously. There were many similar objections against the kinetic theory of heat.[12] More recently, Max Born has based his arguments against the possibility of a return to determinism upon the consistency condition and the assumption which we shall here take for granted, that wave mechanics is incompatible with determinism.

> If any future theory should be deterministic it cannot be a modification of the present one, but must be entirely different. How this should be possible without sacrificing a whole treasure of well established results [i.e., without contradicting highly confirmed physical laws and thereby violating the consistency condition] I leave the determinist to worry about.[13]

Most members of the so-called Copenhagen school of quantum theory would argue in a similar manner. For them the idea of complementarity and the formalism of quantization expressing this idea do not contain any hypothetical element as they are 'uniquely determined by the facts'.[14] Any theory which contradicts this idea is factually inadequate and must be removed. Conversely, an explanation of the idea of complementarity is acceptable only if it either contains this idea, or is at least consistent with it. This is how the consistency condition is used in arguments against theories such as those of Bohm, de Broglie and Vigier.[15]

The use of the consistency condition is not restricted to such general remarks, however, A decisive part of the existing quantum theory *itself*, viz., the projection postulate,[16] is the result of the attempt to give an account of the definiteness of macro objects and macro events that is in accordance with the consistency condition. The influence of the condition of meaning invariance goes even further.

> The Copenhagen-interpretation of the quantum theory [writes Heisenberg[17]] starts from paradox. Any experiment in physics, whether it refers to the phenomena of daily life or to atomic events, is to be described in the terms of classical physics ... *We cannot and should not replace these concepts by any others* [my italics]. Still the application of these concepts is limited by the relations of uncertainty. We must keep in mind this limited range of applicability of the classical concepts while using them, but we cannot and should not try to improve them.

This means that the meaning of the classical terms must remain invariant

[11] E. Mach, *Zwei Aufsätze* (Leipzig, 1912).
[12] For a discussion of these objections, <see> ter Haar's review article, 'Foundations of Statistical Mechanics', *Reviews of Modern Physics*, vol. 27, 1955, pp. 289–338.
[13] M. Born, *Natural Philosophy of Cause and Chance* (Oxford: Clarendon Press, 1949), p. 109.
[14] L. Rosenfeld, 'Misunderstandings about the Foundations of the Quantum Theory', in S. Körner (ed.), *Observation and Interpretation* (London: Butterworth, 1957), p. 42.
[15] <See> the discussions in Körner, *Observation and Interpretation*.
[16] For details and further literature, <see> Section 11 of my paper 'Problems of Microphysics'.
[17] W. Heisenberg, *Physics and Philosophy* (London: Allen and Unwin, 1958), p. 46.

with respect to any future explanation of microphenomena. Microtheories have to be formulated in such a manner that this invariance is guaranteed. The principle of correspondence and the formalism of quantization connected with it were explicitly devised for satisfying this demand. Altogether, the quantum theory seems to be the first theory after the downfall of the Aristotelian physics that has been quite explicitly constructed with an eye both on the consistency condition and the condition of (empirical) meaning invariance. In this respect it is very different indeed from, say, relativity which violates both consistency and meaning invariance with respect to earlier theories. Most of the arguments used for the defence of its customary interpretation also depend on the validity of these two conditions and they will collapse with their removal. An examination of these conditions is therefore very topical and bound deeply to affect present controversies in microphysics. I shall start this investigation by showing that some of the most interesting developments of physical theory in the past have violated both conditions.

3 THESE CONDITIONS NOT INVARIABLY ACCEPTED BY ACTUAL SCIENCE

The case of the consistency condition can be dealt with in a few words: it is well known (and has also been shown in great detail by Duhem)[18] that Newton's theory is inconsistent with Galileo's law of the free fall and with Kepler's laws; that statistical thermodynamics is inconsistent with the second law of the phenomenological theory; that wave optics is inconsistent with geometrical optics; and so on. Note that what is being asserted here is *logical* inconsistency; it may well be that the differences of prediction are too small to be detectable by experiment. Note also that what is being asserted is not the inconsistency of, say, Newton's theory and Galileo's law, but rather the inconsistency of *some consequences* of Newton's theory in the domain of validity of Galileo's law, and Galileo's law. In this last case the situation is especially clear. Galileo's law asserts that the acceleration of the free fall is a constant, whereas application of Newton's theory to the surface of the earth gives an acceleration that is not a constant but *decreases* (although imperceptibly) with the distance from the centre of the earth. Conclusion: if actual scientific procedure is to be the measure of method, then the consistency condition is inadequate.

The case of meaning invariance requires a little more argument, not because it is intrinsically more difficult, but because it seems to be much more closely connected with deep-rooted prejudices. Assume that an explanation is required, in terms of the special theory of relativity, of the

---

[18] P. Duhem, *La Théorie Physique: Son Objet, Sa Structure* (Paris, Marcel Rivière, 1914), chapters IX and X. See also K. R. Popper, 'The Aim of Science', *Ratio*, vol. 1, 1957.

classical conservation of mass in all reactions in a closed system S. If m', m'', m''', $\ldots$, m^i, $\ldots$ are the masses of the parts P', P'', P''', $\ldots$, P^i $\ldots$ of S, then what we want is an explanation of

(1) $$\sum m^i = \text{const.}$$

for all reactions inside S. We see at once that the consistency condition cannot be fulfilled: according to special relativity $\sum m^i$ will vary with the velocities of the parts relative to the co-ordinate system in which the observations are carried out, and the total mass of S will also depend on the relative potential energies of the parts. However, if the velocities and the mutual forces are not too large, then the variation of $\sum m^i$ predicted by relativity will be so small as to be undetectable by experiment. Now let us turn to the *meanings* of the terms in the relativistic law and in the corresponding classical law. The first indication of a possible change of meaning may be seen in the fact that in the classical case the mass of an aggregate of parts equals the sum of the masses of the parts:

$$M\left(\sum P^i\right) = \sum M(P^i)$$

This is not valid in the case of relativity where the relative velocities and the relative potential energies contribute to the mass balance. That the relativistic concept and the classical concept of mass are very different indeed becomes clear if we also consider that the former is a *relation*, involving relative velocities, between an object and a co-ordinate system, whereas the latter is a *property* of the object itself and independent of its behaviour in co-ordinate systems. True, there have been attempts to give a relational analysis even of the classical concept (Mach). None of these attempts, however, leads to the relativistic idea with its velocity dependence on the co-ordinate system, which idea must therefore be added even to a *relational* account of classical mass. The attempt to identify the classical mass with the relativistic rest mass is of no avail either. For although both may have the same numerical value, the one is still dependent on the co-ordinate system chosen (in which it is at rest and has that specific value), whereas the other is not so dependent. We have to conclude, then, that $(m)_c$ and $(m)_r$ mean very different things and that $\left(\sum m^i\right)_c = \text{const.}$ and $\left(\sum m^i\right)_r = \text{const.}$ are very different assertions. This being the case, the derivation from relativity of either equation (1) or of a law that makes slightly different quantitative predictions with $\sum m^i$ used in the classical manner, will be possible only if a further premise is added which establishes a relation between the $(m)_c$ and the $(m)_r$. Such a 'bridge law' – and this is a major point in Nagel's theory of reduction – is a hypothesis

according to which the occurrence of the properties designated by some expression in the premises of the [explanans] is a sufficient, or a necessary and sufficient condition for the occurrence of the properties designated by the expression of the [explanandum].[19]

Applied to the present case this would mean the following: under certain conditions the occurrence of relativistic mass of a given magnitude is accompanied by the occurrence of classical mass of a corresponding magnitude; this assertion is inconsistent with another part of the explanans, viz., the theory of relativity. After all, this theory asserts that there are no invariants which are directly connected with mass measurements and it thereby asserts that '$(m)_c$' does not express real features of physical systems. Thus we inevitably arrive at the conclusion that mass conservation cannot be explained in terms of relativity (or 'reduced' to relativity) without a violation of meaning invariance. And if one retorts, as has been done by some critics of the ideas expressed in the present paper,[20] that meaning invariance is an essential part of both reduction and explanation, then the answer will simply be that equation (1) can neither be explained by, nor reduced to relativity. Whatever the *words* used for describing the situation, the *fact* remains that actual science does not observe the requirement of meaning invariance.

This argument is quite general and is independent of whether the terms whose meaning is under investigation are observable or not. It is therefore stronger than may seem at first sight. There are some empiricists who would admit that the meaning of theoretical terms may be changed in the course of scientific progress. However, not many people are prepared to extend meaning *variance* to observational terms also. The idea motivating this attitude is, roughly, that the meaning of observational terms is uniquely determined by the procedures of observation such as looking, listening, and the like. These procedures remain unaffected by theoretical advance.[21] Hence, observational meanings, too, remain unaffected by theoretical advance. What is overlooked, here, is that the 'logic' of the observational terms is not exhausted by the procedures which are connected with their application 'on the basis of observation'. As will turn out later, it also depends on the more general ideas that determine the 'ontology' (in Quine's sense) of our discourse. These general ideas may change without any change of observational procedures being implied. For example, we may change our ideas about the nature, or the ontological status (property, relation, object, process, etc.) of the colour of a self-luminescent object

---

[19] E. Nagel, 'The Meaning of Reduction', p. 302.

[20] <See> Section 4.7 of M. Scriven's paper 'Explanations, Predictions and Laws', in *Minnesota Studies in the Philosophy of Science, volume III* (Minneapolis: University of Minnesota Press, 1962). Similar objections have been raised by Kraft (Vienna) and Rynin (Berkeley).

[21] For an exposition and criticism of this idea <see> my 'Attempt at a Realistic Interpretation of Experience', *Proceedings of the Aristotelian Society*, vol. 58, 1958, pp. 143–70.

without changing the methods of ascertaining that colour (looking, for example). Clearly, such a change is bound profoundly to influence the meanings of our observational terms.

All this has a decisive bearing upon some contemporary ideas concerning the interpretation of scientific theories. According to these ideas, theoretical terms receive their meanings via correspondence rules which connect them with an observational language *that has been fixed in advance* and independently of the structure of the theory to be interpreted. Now, our above analysis would seem to show that *if we interpret scientific theories in the manner accepted by the scientific community,* then most of these correspondence rules will be either false, or nonsensical. They will be *false* if they *assert* the existence of entities denied by the theory; they will be *nonsensical* if they *presuppose* this existence. Turning the argument around, we can also say that the attempt to interpret the calculus of some theory that has been voided of the meaning assigned to it by the scientific community with the help of the double language system, will lead to a very different theory. Let us again take the theory of relativity as an example: it can be safely assumed that the physical thing language of Carnap, and any similar language that has been suggested as an observation language, is not Lorentz-invariant. The attempt to interpret the *calculus* of relativity on *its* basis therefore cannot lead to the *theory* of relativity as it was understood by Einstein. What we shall obtain will be at the very most *Lorentz's interpretation* with its inherent asymmetries. This undesirable result cannot be evaded by the *demand* to use a different and more adequate observation language. The double language system assumes that theories which are not connected with some observation language do not possess an interpretation. The demand assumes that they do, and asks to choose the observation language most suited to it. It reverses the relation between theory and experience that is characteristic for the double language method of interpretation, which means, it gives up this method. Contemporary empiricism, therefore, has not led to any satisfactory account of the meanings of scientific theories.[22]

What we have shown so far is that the two conditions of Section 2 are frequently violated in the course of scientific practice and especially at

[22] It must be admitted, however, that Einstein's original interpretation of the special theory of relativity is hardly ever used by contemporary physicists. For them the theory of relativity consists of two elements: (1) the Lorentz transformations; and (2) mass-energy equivalence. The Lorentz transformations are interpreted purely formally and are used to make a selection among possible equations. This interpretation does not allow <us> to distinguish between Lorentz's original point of view and the entirely different point of view of Einstein. According to it Einstein achieved a very minor *formal* advance [this is the basis of Whittaker's attempt to 'debunk' Einstein]. It is also very similar to what application of the double language model would yield. Still, an undesirable philosophical procedure is not improved by the support it gets from an undesirable procedure in physics. [The above comment on the contemporary attitude towards relativity was made by E. L. Hill in discussions at the Minnesota Center for the Philosophy of Science.]

periods of scientific revolution. This is not yet a very strong argument. True: there are empirically inclined philosophers who have derived some satisfaction from the assumption that they only make explicit what is implicitly contained in scientific practice. It is therefore quite important to show that scientific practice is not what it is supposed to be by them. Also, strict adherence to meaning invariance and consistency would have made impossible some very decisive advances in physical theory such as the advance from the physics of Aristotle to the physics of Galileo and Newton. However, how do we know (independently of the fact that they do exist, have a certain structure, and are very influential – a circumstance that will have great weight with opportunists only)[23] that the sciences are a desirable phenomenon, that they contribute to the advancement of knowledge, and that their analysis will therefore lead to reasonable methodological demands? And did it not emerge in the last section that meaning invariance and the consistency condition *are* adopted by some scientists? Actual scientific practice, therefore, cannot be our last authority. We have to find out whether consistency and meaning invariance are *desirable* conditions and this quite independently of who accepts and praises them and how many Nobel prizes have been won with their help.[24] Such an investigation will be carried out in the next sections.

4 INHERENT UNREASONABLENESS OF CONSISTENCY CONDITION

Prima facie, the case of the consistency condition can be dealt with in very few words. Consider for that purpose a theory T′ that successfully describes the situation in the domain D′. From this we can infer (a) that T′ agrees with a *finite* number of observations (let their class be F); and (b) that it agrees with these observations inside a margin M of error only.[25] Any alternative that contradicts T′ outside F and inside M is supported by

---

[23] In about 1925 philosophers of science were bold enough to stick to their theses even in those cases where they were inconsistent with actual science. They meant to be *reformers* of science, and not *imitators*. (This point was explicitly made by Mach in his controversy with Planck. Cf. again his *Zwei Aufsätze*.) In the meantime they have become rather tame (or beat) and are much more prepared to change their ideas in accordance with the latest discoveries of the historians, or the latest fashion of the contemporary scientific enterprise. This is very regrettable, indeed, for it considerably decreases the number of the rational critics of the scientific enterprise. And it also seems to give unwanted support to the Hegelian thesis (which is now implicitly held by many historians and philosophers of science) that what exists has a 'logic' of its own and is for that very reason reasonable.

[24] Even the most dogmatic enterprise allows for discoveries (cf. the 'discovery' of so-called 'white Jews' among German physicists during the Nazi period). Hence, before hailing a so-called discovery we must make sure that the system of thought which forms its background is not of a dogmatic kind.

[25] The indefinite character of all observations has been made very clear by Duhem, *La Théorie Physique*, Chap. IX. For an alternative way of dealing with this indefiniteness cf. S. Körner, *Conceptual Thinking* (New York, Dover Publications, 1959).

exactly the same observations and therefore <is> acceptable if T' was acceptable (we shall assume that F are the only observations available). The consistency condition is much less tolerant. It eliminates a theory not because it is in disagreement with the *facts*; it eliminates it because it is in disagreement with *another theory*, with a theory, moreover, whose confirming instances it shares. *It thereby makes the as yet untested part of that theory a measure of validity.* The only difference between such a measure and a more recent theory is age and familiarity. Had the younger theory been there first, then the consistency condition would have worked in its favour. In this respect the effect of the consistency condition is rather similar to the effect of the more traditional methods of transcendental deduction, analysis of essences, phenomenological analysis, linguistic analysis. It contributes to the preservation of the old and familiar not because of any inherent advantage in it – for example, not because it has a better foundation in observation than has the newly suggested alternative, or because it is more elegant – but just because it is old and familiar. This is not the only instance where on closer inspection a rather surprising similarity emerges between modern empiricism and some of the school philosophies it attacks.

Now it seems to me that these brief considerations, although leading to an interesting *tactical* criticism of the consistency condition, do not yet go to the heart of the matter. They show that an alternative of the accepted point of view which shares its confirming instances cannot be *eliminated* by factual reasoning. They do not show that such an alternative is *acceptable*; and even less do they show that it *should be used*. It is bad enough, so a defender of the consistency condition might point out, that the accepted point of view does not possess full empirical support. Adding new theories *of an equally unsatisfactory character* will not improve the situation; nor is there much sense in trying to *replace* the accepted theories by some of their possible alternatives. Such replacement will be no easy matter. A new formalism may have to be learned and familiar problems may have to be calculated in a new way. Textbooks must be rewritten, university curricula readjusted, experimental results reinterpreted. And what will be the result of all the effort? Another theory which, from an empirical point of view, has no advantage whatever over and above the theory it replaces. The only real improvement, so the defender of the consistency condition will continue, derives from the *addition of new facts*. Such new facts will either support the current theories, or they will force us to modify them by indicating precisely where they go wrong. In both cases they will precipitate real progress and not only arbitrary change. The proper procedure must therefore consist in the confrontation of the accepted point of view with as many relevant facts as possible. The exclusion of alternatives is then required for reasons of expediency: their invention not only does not help, but it even hinders progress by absorbing time and manpower that could be devoted to better things. And the function of the consistency condition lies precisely in this. It eliminates such fruitless discussion and it forces the scientist to concentrate on the facts which, after all, are the only acceptable judges of a theory. This is how the practising scientist will defend his concentration on a single theory to the exclusion of all empirically possible alternatives.[26]

It is worthwhile repeating the reasonable core of this argument: theories should not be changed unless there are pressing reasons for doing so. The only pressing reason for changing a theory is disagreement with facts. Discussion of incompatible facts will therefore lead to progress. Discussion of incompatible alternatives will not. Hence, it is sound procedure to increase the number of relevant facts. It is not sound procedure to increase the number of factually adequate, but incompatible alternatives. One might wish to add that formal improvements such as increase of elegance, simplicity, generality and coherence should not be excluded. But once these improvements have been carried out, the collection of facts for the purpose of test seems indeed to be the only thing left to the scientist.

5 RELATIVE AUTONOMY OF FACTS

And this it is – provided these facts *exist, and are available independently of whether or not one considers alternatives to the theory to be tested*. This assumption on which the validity of the argument in the last section depends in a most decisive manner I shall call the assumption of the relative autonomy of facts, or the autonomy principle. It is not asserted by this principle that the discovery and description of facts is independent of *all* theorizing. But it *is* asserted that the facts which belong to the empirical content of some theory are available whether or not one considers alternatives to *this* theory. I am not aware that this very important assumption has ever been explicitly formulated as a separate postulate of the empirical method. However, it is clearly implied in almost all investigations which deal with questions of confirmation and test. All these investigations use a model in which a *single* theory is compared with a class of facts (or observation statements) which are assumed to be 'given' somehow. I submit that this is much too simple a picture of the actual situation. Facts and theories are much more intimately

---

[26] More detailed evidence for the existence of this attitude and for the way in which it influences the development of the sciences may be found in Kuhn's book *The Structure of Scientific Revolutions* (Chicago: University of Chicago Press, 1962). The attitude is extremely common in the contemporary quantum theory. 'Let us enjoy the successful theories we possess and let us not waste our time with contemplating what *would* happen if *other* theories were used' – this seems to be the motto of almost all contemporary physicists (<see> Heisenberg, *Physics and Philosophy*, pp. 56, 144) and philosophers (<see> N. R. Hanson, 'Five Cautions for the Copenhagen Interpretation's Critics', *Philosophy of Science*, vol. 26, 1959, pp. 325–37). It may be traced back to Newton's papers and letters (to Hooke, and Pardies) on the theory of colour. See also footnote 23, above.

connected than is admitted by the autonomy principle. Not only is the description of every single fact dependent on *some* theory (which may, of course, be very different from the theory to be tested). There exist also facts which cannot be unearthed except with the help of alternatives to the theory to be tested, and which become unavailable as soon as such alternatives are excluded. This suggests that the methodological unit to which we must refer when discussing questions of test and empirical content is constituted by a *whole set of partly overlapping, factually adequate, but mutually inconsistent theories.* In the present paper only the barest outlines will be given of such a test model. However, before doing this I want to discuss an example which shows very clearly the function of alternatives in the discovery of facts.

As is well known, the Brownian particle is a perpetual motion machine of the second kind and its existence refutes the phenomenological second law. It therefore belongs to the domain of relevant facts for this law. Now, could this relation between the law and the Brownian particle have been discovered in a *direct* manner, i.e., could it have been discovered by an investigation of the observational consequences of the phenomenological theory that did not make use of an alternative account of heat? This question is readily divided into two: (1) Could the *relevance* of the Brownian particle have been discovered in this manner? (2) Could it have been demonstrated that it actually *refutes* the second law? The answer to the first question is that we do not know. It is impossible to say what would have happened had the kinetic theory not been considered by some physicists. It is my guess, however, that in this case the Brownian particle would have been regarded as an oddity much in the same way in which some of the late Professor Ehrenhaft's astounding effects[27] are regarded as an oddity, and that it would not have been given the decisive position it assumes in contemporary theory. The answer to the second question is simply – No. Consider what the discovery of the inconsistency between the Brownian particle and the second law would have required! It would have required (a) measurement of the exact *motion* of the particle in order to ascertain the changes of its kinetic energy plus the energy spent on overcoming the resistance of the fluid; and (b) it would have required precise measurements of temperature and heat transfer in the surrounding medium in order to ascertain that any loss occurring here was indeed compensated by the increase of the energy of the moving particle and the work done against the fluid. Such measurements are beyond experimental possibilities (<see> R. Fürth, 'Über einige Beziehungen zwischen klassicher Statistik und

---

[27] Having witnessed these effects under a great variety of conditions, I am much more reluctant to regard them as mere curiosities than is the scientific community of today. <See> also my edition of Ehrenhaft's lectures, *Einzelne Magnetische Nord- und Südpole und deren Auswirkung in den Naturwissenschaften* (Vienna, 1947).

Quantenmechanik', *Zeitschrift für Physik*, vol. 81, 1933, pp. 143–62). Neither is it possible to make precise measurements of the heat transfer; nor can the path of the particle be investigated with the desired precision. Hence a 'direct' refutation of the second law that considers only the phenomenological theory and the 'facts' of Brownian motion is impossible. And, as is well known, the actual refutation was brought about in a very different manner. It was brought about via the kinetic theory and Einstein's utilization of it in the calculation of the statistical properties of the Brownian motion.[28] In the course of this procedure the phenomenological theory (T') was incorporated into the wider context of statistical physics (T) *in such a manner that the consistency condition was violated*; and *then* a crucial experiment was staged (investigations of Svedberg and Perrin).

It seems to me that this example is typical for the relation between fairly general theories, or points of view, and 'the facts'. Both the relevance and the refuting character of many very decisive facts can be established only with the help of other theories which, although factually adequate, are yet not in agreement with the view to be tested. This being the case, the production of such refuting facts may have to be preceded by the invention and articulation of alternatives to that view. Empiricism demands that the empirical content of whatever knowledge we possess be increased as much as possible. Hence *the invention of alternatives in addition to the view that stands in the centre of discussion constitutes an essential part of the empirical method.* Conversely, the fact that the consistency condition eliminates alternatives now shows it to be in disagreement with empiricism and not only with scientific practice. By excluding valuable tests it decreases the empirical content of the theories which are permitted to remain (and which, as we have indicated above, will usually be the theories which have been there first); and it especially decreases the number of those facts which could show their limitations. This last result of a determined application of the consistency condition is of very topical interest. It may well be that the

---

[28] For these investigations, <see> A. Einstein, *Investigations on the Theory of the Brownian Motion* (New York: Dover, 1956), which contains all the relevant papers by Einstein and an exhaustive bibliography by R. Fürth. For the experimental work, <see> J. Perrin, *Die Atome* (Leipzig, 1920). For the relation between the phenomenological theory and the kinetic theory, <see> also M. v. Smoluchowski, 'Experimentell nachwiesbare, der üblichen Thermodynamik widersprechende Molekularphänomene', *Physikalische Zeitschrift*, vol. 13, 1912, p. 1069, and K. R. Popper, 'Irreversibility; or, Entropy since 1905', *British Journal for the Philosophy of Science*, vol. 8, 1957, pp. 151–5. Despite Einstein's epoch-making discoveries and von Smoluchowski's splendid presentation of their effect (for the latter <see> also *Œuvres de Marie Smoluchowski, volume 2* (Cracovie, 1927), pp. 226 ff., 316 ff., 462 ff. and 530 ff.) the present situation in thermodynamics is extremely unclear, especially in view of the continued presence of the ideas of reduction which we criticized in the text above. To be more specific, it is frequently attempted to determine the entropy balance of a complex *statistical* process by reference to the (refuted) *phenomenological* law after which procedure fluctuations are superimposed in a most artificial fashion. For details cf. Popper, 'Irreversibility'.

refutation of the quantum-mechanical uncertainties presupposes just such an incorporation of the present theory into a wider context which is no longer in accordance with the idea of complementarity and which therefore suggests new and decisive experiments. And it may also be that the insistence, on the part of the majority of contemporary physicists, on the consistency condition will, if successful, forever protect these uncertainties from refutation. This is how modern empiricism may finally lead to a situation where a certain point of view petrifies into dogma by being, in the name of experience, completely removed from any conceivable criticism.

6 THE SELF-DECEPTION INVOLVED IN ALL UNIFORMITY

It is worthwhile to examine this apparently empirical defence of a dogmatic point of view in somewhat greater detail. Assume that physicists have adopted, either consciously or unconsciously, the idea of the uniqueness of complementarity and that they therefore elaborate the orthodox point of view and refuse to consider alternatives. In the beginning such a procedure may be quite harmless. After all, a man can do only so many things at a time and it is better when he pursues a theory in which he is interested rather than a theory he finds boring. Now assume that the pursuit of the theory he chose has led to successes and that the theory has explained in a satisfactory manner circumstances that had been unintelligible for quite some time. This gives empirical support to an idea which to start with seemed to possess only this advantage: it was interesting and intriguing. The concentration upon the theory will now be reinforced, the attitude towards alternatives will become less tolerant. Now if it is true, as has been argued in the last section, that many facts become available only with the help of such alternatives, then the refusal to consider them *will result in the elimination of potentially refuting facts*. More especially, it will eliminate facts whose discovery would show the complete and irreparable inadequacy of the theory.[29] Such facts having been made inaccessible, the theory will appear to be free from blemish and it will seem that 'all evidence points with merciless definiteness in the ... direction ... [that] all the processes involving ... unknown interactions conform to the fundamental quantum law' (Rosenfeld in Körner (ed.), *Observation and Interpretation*. This will

---

[29] The quantum theory can be adapted to a great many difficulties. It is an open theory in the sense that apparent inadequacies can be accounted for in an *ad hoc* manner, by *adding* suitable operators, or elements in the Hamiltonian, rather than by recasting the whole structure. A refutation of its basic formalism (i.e., of the formalism of quantization, and of non-commuting operators in a Hilbert space or a reasonable extension of it) would therefore demand proof to the effect that *there is no conceivable adjustment of the Hamiltonian, or of the operators used* which makes the theory conform to a given fact. It is clear that such a general statement can only be provided by an *alternative theory* which of course must be detailed enough to allow for independent and crucial tests.

further reinforce the belief in the uniqueness of the current theory and in the complete futility of any account that proceeds in a different manner. Being now very firmly convinced that there is only one good microphysics, the physicists will try to explain even adverse facts in its terms, and they will not mind when such explanations are sometimes a little clumsy. By now the success of the theory has become public news. Popular science books (and this includes a good many books on the philosophy of science) will spread the basic postulates of the theory; applications will be made in distant fields. More than ever the theory will appear to possess tremendous empirical support. The chances for the consideration of alternatives are now very slight indeed. The final success of the fundamental assumptions of the quantum theory and of the idea of complementarity will seem to be assured.

At the same time it is evident, on the basis of the considerations in the last section, that this appearance of success *cannot in the least be regarded as a sign of truth and correspondence with nature.* Quite the contrary, the suspicion arises that the absence of major difficulties is a result of the decrease of empirical content brought about by the elimination of alternatives, and of facts that can be discovered with the help of these alternatives only. In other words, *the suspicion arises that this alleged success is due to the fact that in the process of application to new domains the theory has been turned into a metaphysical system.* Such a system will of course be very 'successful' not, however, because it agrees so well with the facts, but because no facts have been specified that would constitute a test and because some such facts have even been removed. Its 'success' *is entirely man made.* It was decided to stick to some ideas and the result was, quite naturally, the survival of these ideas. If now the initial decision is forgotten, or made only implicitly, then the survival will seem to constitute independent support, it will reinforce the decision, or turn it into an explicit one, and in this way close the circle. This is how empirical 'evidence' may be *created* by a procedure which quotes as its justification the very same evidence it has produced in the first place.

At this point an 'empirical' theory of the kind described (and let us always remember that the basic principles of the present quantum theory and especially the idea of complementarity are uncomfortably close to forming such a theory) becomes almost indistinguishable from a myth. In order to realize this, we need only consider that on account of its all-pervasive character a myth such as the myth of witchcraft and of demonic possession will possess a high degree of confirmation on the basis of observation. Such a myth has been taught for a long time; its content is enforced by fear, prejudice and ignorance as well as by a jealous and cruel priesthood. It penetrates the most common idiom, infects all modes of thinking and many decisions which mean a great deal in human life. It provides models for the explanation of any conceivable event, conceivable, that is, for those who have accepted it.[30] This being the case, its key terms will be fixed in an unambiguous manner and the idea (which may have led to such a procedure in the first place) that they are copies of unchanging entities and that change of meaning, if it should happen, is due to human mistake – this idea will now be very plausible. Such plausibility reinforces all the manoeuvres which are used for the preservation of the myth (elimination of opponents included). The conceptual apparatus of the theory and the emotions connected with its application having penetrated all means of communication, all actions, and indeed the whole life of the community, such methods as transcendental deduction, analysis of usage, phenomenological analysis which are means for further solidifying the myth will be extremely successful (which shows, by the way, that all these methods which have been the trademark of various philosophical schools old and new, have one thing in common: they tend to *preserve* the *status quo* of the intellectual life).[31] Observational results, too, will speak in favour of the theory as they are formulated in its terms. It will seem that at last the truth has been arrived at. At the same time it is evident that all contact with the world has been lost and that the stability achieved, the semblance of absolute truth, *is nothing but the result of an absolute conformism.*[32] For how can we possibly test, or improve upon, the truth of a theory if it is built in such a manner that any conceivable event can be described, and explained, in terms of its principles? The *only* way of investigating such all-embracing principles is to compare them with a different set of *equally all-embracing* principles – but this way has been excluded from the very beginning. The myth is therefore of no objective relevance, it continues to exist solely as the result of the effort of the community of believers and of their leaders, be these now priests or Nobel prize winners. *Its 'success' is entirely man made.* This, I think, is the most decisive argument against any method that encourages uniformity, be it now empirical or not. Any such method is in the last resort a method of deception. It enforces an unenlightened

---

[30] For a very detailed description of a once very influential myth, <see> H. C. Lea, *Materials Towards a History of Witchcraft*, 3 vols. (Philadelphia: University of Pennsylvania Press, 1939), as well as *Malleus Malleficarum* (London: John Rodker, 1928), translated by Montague Summers (who, by the way, counts it 'among the most important, wisest [sic!], and weightiest books of the world').

[31] Quite clearly, analysis of usage, to take only one example, presupposes certain regularities concerning this usage. The more people differ in their fundamental ideas, the more difficult it will be to uncover such regularities. Hence, analysis of usage will work best in a closed society that is firmly held together by a powerful myth such as was the philosophy in the Oxford of about ten years ago.

[32] Schizophrenics very often hold beliefs which are as rigid, all-pervasive, and unconnected with reality, as are the best dogmatic philosophies. Only such beliefs come to them naturally whereas a professor may sometimes spend his whole life in attempting to find arguments which create a similar state of mind.

conformism, and speaks of truth; it leads to a deterioration of intellectual capabilities, of the power of imagination, and speaks of deep insight; it destroys the most precious gift of the young, their tremendous power of imagination, and speaks of education.

To sum up: *Unanimity of opinion may be fitting for a church, for the frightened victims of some (ancient, or modern) myth, or for the weak and willing followers of some tyrant; variety of opinion is a feature necessary for objective knowledge; and a method that encourages variety is also the only method that is compatible with a humanitarian outlook.* To the extent to which the consistency condition (and, as will emerge, the condition of meaning invariance) delimits variety, it contains a theological element (which lies, of course, in the worship of 'facts' so characteristic for nearly all empiricism).

7 INHERENT UNREASONABLENESS OF MEANING INVARIANCE

What we have achieved so far has immediate application to the question whether the meaning of certain key terms should be kept unchanged in the course of the development and improvement of our knowledge. After all, the meaning of every term we use depends upon the theoretical context in which it occurs. Hence, if we consider two contexts with basic principles which either contradict each other, or which lead to inconsistent consequences in certain domains, it is to be expected that some terms of the first context will not occur in the second context with exactly the same meaning. Moreover, if our methodology demands the use of mutually inconsistent, partly overlapping, and empirically adequate theories, then it thereby also demands the use of conceptual systems which are mutually *irreducible* (their primitives cannot be connected by bridge laws which are meaningful *and* factually correct) and it demands that meanings of terms be left elastic and that no binding commitment be made to a certain set of concepts.

It is very important to realize that such a tolerant attitude towards meanings, or such a change of meaning in cases where one of the competing conceptual systems has to be abandoned need not be the result of directly accessible observational difficulties. The law of inertia of the so-called *impetus theory* of the later Middle Ages[33] and Newton's own law of inertia are in perfect quantitative agreement: both assert that an object that is not under the influence of any outer force will proceed along a straight line with constant speed. Yet despite this fact, the adoption of Newton's theory entails a conceptual revision that forces us to abandon the inertial law of the impetus theory, not because it is quantitatively incorrect but *because it achieves the correct predictions with the help of inadequate concepts.* The law

---

[33] For details and further references, <see> Section 6 of my 'Explanation, Reduction, and Empiricism'.

asserts that the *impetus* of an object that is beyond the reach of outer forces remains constant.[34] The impetus is interpreted as an inner *force* which pushes the object along. Within the impetus theory such a force is quite conceivable as it is assumed here that forces determine *velocities* rather than accelerations. The concept of impetus is therefore formed in accordance with a law (forces determine velocities) and this law is inconsistent with the laws of Newton's theory and must be abandoned as soon as the latter is adopted. This is how the progress of our knowledge may lead to conceptual revisions for which no direct observational reasons are available. The occurrence of such changes quite obviously refutes the contention of some philosophers that the invariance of *usage* in the trivial and uninteresting contexts of the private lives of not too intelligent and inquisitive people indicates invariance of *meaning* and the superficiality of all scientific changes. It is also a very decisive objection against any crudely operationalistic account of both observable terms and theoretical terms.

What we have said applies even to singular statements of observation. Statements which are empirically adequate, and which are the result of observation (such as 'here is a table') may have to be reinterpreted, not because it has been found that they do not adequately express what is seen, heard, felt, but because of some changes in sometimes very remote parts of the conceptual scheme to which they belong. Witchcraft is again a very good example. Numerous eyewitnesses claim that they have actually *seen* the devil or *experienced* demonic influence. There is no reason to suspect that they were lying. Nor is there any reason to assume that they were sloppy observers, for the phenomena leading to the belief in demonic influence are so obvious that a mistake is hardly possible (possession; split personality; loss of personality; hearing voices; etc.). These phenomena are well known today.[35] In the conceptual scheme that was the one generally accepted in the fifteenth and sixteenth centuries, the only way of describing them, or at least the way that seemed to express them most adequately, was by reference to demonic influences. Large parts of this conceptual scheme were changed for philosophical reasons and also under the influence of the evidence accumulated by the sciences. Descartes's materialism played a very decisive role in discrediting the belief in spatially localizable spirits. The language of demonic influences was no part of the new conceptual scheme that was created in this manner. It was for this reason that a reformulation was needed, and a reinterpretation of even the most

---

[34] We assume here that a dynamical rather than a kinematic characterization of motion has been adopted. For a more detailed analysis <see> again the paper referred to in the previous footnote.

[35] For very vivid examples, <see> K. Jaspers, *Allgemeine Psychopathologie* (Berlin: Springer-Verlag, 1959), pp. 75–123.

common 'observational' statements. Combining this example with the remarks at the beginning of the present section, we now realize that according to the method of classes of alternative theories a lenient attitude must be taken with respect to the meanings of all the terms we use. We must not attach too great an importance to 'what we mean' by a phrase, and we must be prepared to change whatever little we have said concerning this meaning as soon as the need arises. Too great concern with meanings can only lead to dogmatism and sterility. Flexibility, and even sloppiness in semantical matters is a prerequisite of scientific progress.[36]

8 SOME CONSEQUENCES

Three consequences of the results so far obtained deserve a more detailed discussion. The first consequence is an evaluation of *metaphysics* which differs significantly from the standard empirical attitude. As is well known, there are empiricists who demand that science start from observable facts and proceed by generalization, and who refuse the admittance of metaphysical ideas at any point of this procedure. For them, only a system of thought that has been built up in a purely inductive fashion can claim to be genuine knowledge. Theories which are partly metaphysical, or 'hypothetical', are suspect, and are best not used at all. This attitude has been formulated most clearly by Newton[37] in his reply to Pardies' second letter concerning the theory of colours:

> if the possibility of hypotheses is to be the test of truth and reality of things, I see not how certainty can be obtained in any science; since numerous hypotheses may be devised, which shall seem to overcome new difficulties.

This radical position, which clearly depends on the demand for a theoretical monism, is no longer as popular as it used to be. It is now granted that metaphysical considerations may be of importance when the task is to *invent* a new physical theory; such invention, so it is admitted, is a more or less irrational act containing the most diverse components. Some of these components are, and perhaps must be, metaphysical ideas. However, it is also pointed out that as soon as the theory has been developed in a formally satisfactory fashion and has received sufficient confirmation to be regarded as empirically successful, it is pointed out that in the very same moment it can *and must* forget its metaphysical past; metaphysical speculation must *now* be replaced by empirical argument.

---

[36] Mae West is by far preferable to the precisionists: 'I ain't afraid of pushin' grammar around so long as it sounds good' (*Goodness Had Nothing to do With It* (New York, 1959), p. 19).

[37] I. B. Cohen (ed.), *Isaac Newton's Papers and Letters on Natural Philosophy* (Cambridge, MA: Harvard University Press, 1958), p. 106.

> On the one side I would like to emphasize [writes Ernst Mach on this point][38] that *every and any* idea is admissible as a means for research, provided it is helpful; still, it must be pointed out, on the other side, that it is very necessary from time to time to free the presentation of the *results* of research from all inessential additions.

This means that empirical considerations are still given the upper hand over metaphysical reasoning. Especially in the case of an inconsistency between metaphysics and some highly confirmed empirical theory it will be decided, *as a matter of course*, that the theory or the result of observation must stay, and that the metaphysical system must go. A very simple example is the way in which materialism is being judged by some of its opponents. For a materialist the world consists of material particles moving in space, of collections of such particles. Sensations, as introspected by human beings, do not look like collections of particles and their observed existence is therefore assumed to refute and thereby to remove the metaphysical doctrine of materialism. Another example which I have analysed in 'Problems of Microphysics' is the attempt to eliminate certain very general ideas concerning the nature of micro-entities on the basis of the remark that they are inconsistent 'with an immense body of experience' and that 'to object to a lesson of experience by appealing to metaphysical preconceptions is unscientific'.[39]

The methodology developed in the present paper leads to a very different evaluation of metaphysics. Metaphysical systems are scientific theories in their most primitive stage. If they *contradict* a well-confirmed point of view, then this indicates their usefulness as an alternative to this point of view. Alternatives are needed for the purpose of criticism. Hence, metaphysical systems which contradict observational results or well-confirmed theories *are most welcome* starting points of such criticism. Far from being misfired attempts at anticipating, or circumventing, empirical research which were deservedly exposed by a reference to experience, they are the only means at our disposal for examining those parts of our knowledge which have already become observational and which are therefore inaccessible to a criticism 'on the basis of observation'.

A second consequence is that a new attitude has to be adopted with respect to the *problem of induction*. This problem consists in the question of what justification there is for asserting the truth of a statement S given the truth of another statement, S', whose content is smaller than the content of S. It may be taken for granted that those who want to justify the truth of S also assume that after the justification the truth of S will be *known*. Knowledge to the effect that S implies the *stability* of S (we must not

---

[38] 'Der Gegensatz zwischen der mechanischen und der phänomenologischen Physik', *Wärmelehre* (Leipzig, 1896), pp. 362 f.
[39] L. Rosenfeld, 'Misunderstandings', p. 42.

change, remove, criticize, what we know to be true). The method we are discussing at the present moment cannot allow such stability. It follows that the problem of induction, at least in some of its formulations, is a problem whose solution leads to undesirable results. It may therefore be properly termed a pseudo problem.

The third consequence, which is more specific, is that *arguments from synonymy* (or from co-extensionality), far from being that measure of adequacy as which they are usually introduced, are liable severely to impede the progress of knowledge. Arguments from synonymy judge a theory or a point of view not by its capability to mimic the world but rather by its capability to mimic the descriptive terms of another point of view which for some reason is received favourably. Thus for example, the attempt to give a materialistic, or else a purely physiological, account of human beings is criticized on the grounds that materialism, or physiology, cannot provide synonyms for 'mind', 'pain', 'seeing red', 'thinking of Vienna', in the sense in which these terms are used either in ordinary English (provided there is a well-established usage concerning these terms, a matter which I doubt) or in some more esoteric mentalistic idiom. Clearly, such criticism silently assumes the principle of meaning invariance, that is, it assumes that the meanings of at least some fundamental terms must remain unchanged in the course of the progress of our knowledge. It cannot therefore be accepted as valid.[40]

However, we can, and must, go still further. The ideas which we have developed above are strong enough not only to *reject* the demand for synonymy, wherever it is raised, but also to *support* the demand for irreducibility (in the sense in which this notion was used at the beginning of Section 7). The reason is that irreducibility is a presupposition of high critical ability on the part of the point of view shown to be irreducible. An outer indication of such irreducibility which is quite striking in the case of an attack upon commonly accepted ideas is the feeling of *absurdity*: we deem absurd what goes counter to well-established linguistic habits. The absence, from a newly introduced set of ideas of synonymy relations connecting it with parts of the accepted point of view; the feeling of absurdity therefore indicates that the new ideas are fit for the purpose of criticism, i.e., that they are fit for either leading to a strong *confirmation* of the earlier theories, or else to a very revolutionary *discovery*: absence of synonymy, clash of meanings, absurdity are desirable. Presence of synonymy, intuitive appeal, agreement with customary modes of speech, far from being *the* philosophical virtue, indicates that not much progress has

---

[40] For details concerning the mind–body problem, <see> my 'Materialism and the Mind–Body Problem', *The Review of Metaphysics*, vol. 17, 1963.

been made and that the business of investigating what is commonly accepted *has not even started.*

9 HOW TO BE A GOOD EMPIRICIST

The final reply to the question put in the title is therefore as follows. A good empiricist will not rest content with the theory that is in the centre of attention and with those tests of the theory which can be carried out in a direct manner. Knowing that the most fundamental and the most general criticism is the criticism produced with the help of alternatives, he will try to invent such alternatives.[41] It is, of course, impossible at once to produce a theory that is formally comparable to the main point of view and that leads to equally many predictions. His first step will therefore be the formulation of fairly general assumptions which are not yet directly connected with observations; this means that his first step will be the invention of a new *metaphysics*. This metaphysics must then be elaborated in sufficient detail in order to be able to compete with the theory to be investigated as regards generality, details of prediction, precision of formulation.[42] We may sum up both activities by saying that a good empiricist must be a critical metaphysician. Elimination of all metaphysics, far from increasing the empirical content of the remaining theories, is liable to turn these theories into dogmas. The consideration of alternatives together with the attempt to criticize each of them in the light of experience also leads to an attitude where meanings do not play a very important role and where arguments are based upon assumptions of fact rather than analysis of (archaic, although perhaps very precise) meanings. The effect of such an attitude upon the development of human capabilities should not be underestimated either. Where speculation and invention of alternatives is encouraged, bright ideas are liable to occur in great number and such ideas may then lead to a change of even the most 'fundamental' parts of our knowledge, i.e., they may lead to a change of assumptions which either are so close to observation that their truth seems to be dictated by 'the facts', or which are so close to common prejudice that they seem to be 'obvious', and their negation 'absurd'. In such a situation it will be realized that neither 'facts' nor abstract ideas can ever be used for defending certain principles come what may. Wherever facts play a role in such a dogmatic defence, we shall have to suspect foul play (see the opening quotation) – the foul play of those who try to turn good science into bad, because unchangeable, metaphysics. In the last resort, therefore, being a good empiricist means

---

[41] In my paper 'Realism and Instrumentalism' I have tried to show that this is precisely the method which has brought about such spectacular advances of knowledge as the Copernican Revolution, the transition to relativity and to quantum theory.

[42] <See> Section 13 of my 'Realism and Instrumentalism'.

being critical, and basing one's criticism not just on an abstract principle of scepticism but upon *concrete suggestions* which indicate in every single case how the accepted point of view might be further tested and further investigated and which thereby prepare the next step in the development of our knowledge.

4

Outline of a pluralistic theory of knowledge and action

HISTORICAL SKETCH

It is a commonplace – and a very ancient one – that man, when growing up, must put aside childish things and must adapt to reality. The dreams of his childhood, the aspirations of his youthful years, the many fanciful ways of looking at the world which made it such an interesting and mysterious place – all these must be abandoned if the aim is to achieve knowledge and a mastery of nature. The way to knowledge and to the conquest of nature consists in increasingly restricting the range of possible ideas until a close fit is established between behaviour and thinking, on the one side, and 'reality', on the other. The aim is reached as soon as a single point of view is established beyond doubt as the one correct picture of the world.

It is interesting to see how many different phases in the history of mankind are united by this commonplace idea. The initiation rites that in a primitive society accentuate the transition to manhood all have the same consequence: they make it almost impossible for the initiate ever again to think along lines essentially different from the ideology of the tribe. Plato, who was well aware of the psychological mechanisms creating a state of firm and unwavering belief, gives an extremely perceptive account of such rites and backs them by arguments to show the absolute truth of the story that *he* would have liked to be believed forever. This double machinery of psychological manipulation and philosophical argument was then developed to perfection by the Church Fathers. Never again was there to be such a deep understanding of human nature; and never again was this understanding used with such fatal effect for the physical and conceptual propagation of ideologies. The new philosophy of the sixteenth and seventeenth centuries differs from its ancestor in only two respects. The content of the doctrines defended differs from that of the preceding ideologies. And the psychological manipulation is left to each individual – it is not institutionalized. Otherwise, the situation is exactly the same: there is only one correct point of view; the correct philosophical method aims at proving its truth; and the correct psychological procedure aims at establishing unanimity as well as steadfastness in the pursuit of truth. There is no doubt at all that the founders of modern philosophy, Descartes and Bacon, were interested in the psychology of belief. Both develop a theory of idols and try to explain why man is so frequently deceived. Both devise methods for undeceiving him. And both recognize that the transition to the new philosophy involves perhaps a rather lengthy period of training that creates a mind capable of understanding the arguments and prepared to cling unwaveringly to their results.

The characteristic just mentioned applies even to the founders of modern science. Galileo especially recognizes the need to prepare the mind of the reader so that he will be able to understand the new astronomy and to remain loyal to it even in the face of difficulties. But while Bacon and Descartes are quite explicit about their enterprise and while they oppose common sense from the outset, Galileo uses parts of common sense and the psychological hold it has upon the individual to destroy the rest. Newton, who is quite different in this respect, restricts himself to experiment and philosophical argument; but his theories soon become the basis for new and institutionalized means of creating unanimity. This is how modern science, or 'mature' science (Kuhn), comes into being. It is no exaggeration to say that it shares many properties with the primitive ideologies outlined earlier. There is, however, one important difference.

The basic doctrine of a closed society determines all the aspects of life in this society. The arts, or those activities which we today characterize as artistic, have the function of either creating or reinforcing the belief in the basic myth. Dances, burial ceremonies, statutes, sacred buildings – all these things are repetitions, in different media, of the basic myth. It is therefore possible, on the one hand, to give a rational explanation of their structure and of the particular rules that the artists use at their trade. On the other hand, there is not a single avenue left open to those who might want to think along different lines. Even dreams conform to the basic pattern. The restriction of the individual is complete.

It is different with the more 'modern' manifestations of conformity. While the modern theories of knowledge are still largely hostile to a proliferation of ideas and while contemporary philosophers and scientists (with an increasing number of exceptions, to be sure) regard it as their main task to delimit the number of alternative views, such a restriction is no longer demanded in the arts. We have here – side by side with more conservative theories – the principle that it is the task of the artist to express himself, and to follow up any idea, any emotion, however peculiar, that might strike his fancy. The dreams of childhood and the aspirations of youth are no longer pushed aside. One may retain them. But although externalized they remain dreams as they are confined to the private sphere of a single individual, or of a school. They are not allowed to develop in the face of resistance, and they do not contribute to the understanding of the material universe in which we live. The impression of mystery that they convey is no guide whatever in these matters. Is this all that can be expected from them? Or is it perhaps possible to ascribe to them a more substantial function? Is it possible to retain what one might call the freedom of artistic creation and yet to utilize it in the improvement of our knowledge? This is the question to which I want to address myself.

REASONS FOR HOPE

There are various reasons why one may hope that this question has an affirmative answer. First of all, let us remember that the development of animal species is the result of a process of proliferation that goes on even if the existing species happen to be well adapted to their surroundings. The limits of their ability and the limits of their degree of adaptation are usually revealed only by the succeeding forms. The criterion is the survival of the variants in situations in which their predecessors fail. Many thinkers have assumed that the same process of proliferation might be useful also in the case of theories (ideologies or plans for the anticipation or the formation of the future). They have pointed out – and this brings me to the second hopeful point – that the faults of the apparently most perfect theory can often be discovered only with the help of alternatives that, while retaining and explaining its success, also explain why it must nevertheless be abandoned. The most striking demonstration of this feature of our knowledge is found in the courtroom procedure. Who does not remember trials where the point of view of the prosecution is supported by the evidence to an extent that makes doubt not only impossible but simply irrational? And who does not remember his surprise at the way in which a clever and resourceful counsel for the defence can develop an alternative interpretation which first provides reasons for seeing the evidence differently and then shows that the alternative is supported by the newly arranged evidence as firmly as was the original view of the prosecution? The starting point of such a procedure is always an idea that is inconsistent with the theory to be criticized and that is therefore initially incompatible with evidence of the most convincing kind. The starting point is therefore what one might call a foolish and absurd conjecture. But such a foolish and absurd conjecture may make one look at the evidence in a new and different light and may lead to the discovery of facts that are fatal for the 'well-established' position.

It is the glory of the judiciary system of most civilized countries that this possibility is fully recognized and is made part of the legal procedure: there must be counsel for the defence even in the most 'obvious' cases. He must be able to challenge not only the legal aspects of the proceedings but any point, even if it is made by the highest 'authority' and by the most qualified 'expert'. We also see how in many trials the assurance of the expert turns out to be without foundation and that he may not know what he is talking about. And, mind you, these faults are uncovered, not by a still bigger expert, *but by a clever layman*, by a lawyer who has taken the trouble to study the point in question for two or three weeks and who has the advantage of not being bound by the prejudices of some profession.

This now brings me to the final point. The restrictions that the guardians of knowledge – be they scientists or philosophers – want to impose upon us are usually defined by the latest fashion, by the most popular theory in their own field. They are defined by what some rather clever men have arrived at after long study and patient investigation. They merit serious attention. But these restrictions, this concentration upon a narrow domain of theories indicates also that the scientists have come to the end of their rope, that they can no longer think of any decisive objection (or any decisive reason for defending an alternative) and that they have therefore, for the time being, agreed to accept a single point of view to the exclusion of everything else. Of course, the situation is hardly ever presented in that way. Instead of admitting that their ingenuity has given out and that they are no longer able to advance knowledge, scientists are usually in the habit of saying that they have finally arrived at the truth. But we, looking at the situation from the point of view of a lawyer dealing with an obstinate and conceited expert, cannot permit ourselves to be so easily impressed. A brief look at certain features of what some people are pleased to call the 'scientific method' shows that this attitude is supported not only by general considerations, such as those just given, but also by a more detailed study.

ARGUMENTS FOR PROLIFERATION

It would be imprudent to give up a theory that either is inconsistent with observational results or suffers from internal difficulties. Theories can be developed and improved, and their relation to observation is also capable of modification. It took considerable time until the relation of the kinetic theory to the 'fact' of irreversibility was properly understood, and research in this direction still proceeds. Moreover, it would be a complete surprise if it turned out that all the available experimental results support a certain theory, even if the theory were true. Different observers using different experimental equipment and different methods of interpretation introduce idiosyncrasies and errors of their own, and it takes a long time until all these differences are brought to a common denominator. Considerations like these make us accept a *principle of tenacity*, which suggests, first, that we select from a number of theories the one that has the most attractive features and that promises to lead to the most fruitful results; and, second, that we stick to this theory despite considerable difficulties.

After tenacity is accepted, a theory T can no longer be removed by discordant experiments. One might now be inclined to specify a limit of disagreement beyond which one is not prepared to go. But it is not easy to see how such a limit can be fixed in a non-arbitrary fashion. The most overwhelming difficulties have been overcome, and minor disturbances have proved fatal (just compare the initial difficulties of the heliocentric system with the initial attitude toward the experiment of Michelson and Morley). But it *is* rational to withdraw T if there exists another theory T' that accentuates the difficulties of T (and which is therefore inconsistent with T) while at the same time promising means for their removal and opening up new avenues of research. In this case the principle of tenacity itself urges us to remove T. Such a method of refutation, of course, works only if one is permitted to consider theories inconsistent with T: alternatives to T. The result is that a science that is prepared to develop its theories despite difficulties needs a *principle of proliferation* for the effective criticism of the tenaciously held theories.

So far, we have assumed that the facts endangering T are already available, and we have asked how T can be eliminated if one is prepared to retain it despite their existence. But the situation is frequently much more complex than that. There may exist facts that endanger T but that can be revealed with the help of alternatives only. Assume, for that purpose, that T entails C, that C' is what actually happens, that C', but not C, triggers a macroscopic process M that can be seen by all, and assume further that C and C' are indistinguishable not only because our measuring instruments are too crude but because the laws of nature prohibit the distinction by any physical means. In this case M refutes T, but we can never ascertain that this is so. Only the Good Lord, who stands above all laws of nature and is not bound by them, is able to point out that T is refuted by M – unless the erring mortals are allowed to proliferate and to invent alternatives to T. For if one of these alternatives, say T', predicts C' and the connection between C' and M, and if it approximately repeats the successful predictions S of T and makes in addition some other predictions A, then we shall trust T' more than T and accept the assertion, following from it, that T has been refuted by M. In this case the alternative has not just accentuated an already existing difficulty; it has actually created it. In the light of this possibility, the use of alternatives is recommended even if the theory that stands in the centre of the attention should happen to be without blemish. The refutation of the phenomenological second law of thermodynamics is a case in point.[1]

There is another reason in favour of proliferation which is even more subtle and which has been put forth, quite recently, by Dr Imre Lakatos.[2]

[1] For a more detailed discussion, see section VI of my 'Problems of Empiricism' in Robert G. Colodny (ed.), *Beyond the Edge of Certainty* (Englewood Cliffs, NJ: Prentice-Hall, 1965).

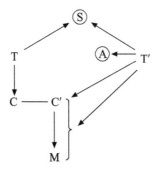

This reason is concerned with what one might call the metaphysical components of observation.

Observational terms, as <sic> it has been frequently assumed, can be defined without reference to theories, simply by exhibiting the phenomena to which they refer. On the other hand, it was also pointed out that observations are always guided by more general ideas. Kant especially emphasized that experience, as conceived by scientists, contains theoretical elements and that observational reports lacking these elements are not admitted into the body of scientific knowledge. It is now generally agreed that a sense-datum language is useless for the purpose of science and that useful observation reports must go beyond what is immediately seen. However, the sense-datum position is still far from abandoned, as is revealed by the tendency to *minimize* the additional, hypothetical element. It is still believed that actual observational reports are hypothetical only to a very small degree.

It is not difficult to show that this assumption is erroneous. Every ordinary statement about medium-sized objects such as 'this table is brown' or, to use more technical terms, every statement of what Rudolf Carnap has called the physical-thing-language contains the idea of observer independence (which entails, among other things, that a fast motion of the observer in the neighbourhood of Sirius will leave tables unaffected). This idea is not only highly hypothetical but also metaphysical (as was already noted by Berkeley); it is metaphysical because it is not possible to specify an experimental result that would endanger it and that might force us to give it up. Ordinary observation statements, therefore, have metaphysical components.

Now Dr Lakatos has shown that one can arrive at the same result in a different and much more interesting manner. Consider a theory *T*.

According to Karl Popper, whose procedure most adequately reflects what is going on in the sciences, T is scientific only if it has potential falsifiers, that is, only if there exist observational statements S_i such that S_i & T is a contradiction (we omit mentioning the conditions that must be imposed in order to eliminate trivial cases). Now in determining the truth value of the S_i, one usually refers to auxiliary theories T' (the test of Newton's celestial mechanics involves optical theory, theory of elasticity, physiology, chemistry, and so on). These auxiliary theories help us to test the S_i, and they also have an influence upon the terms of S_i. It is clear that the strength of the tests of T which are provided by the S_i will be the greater as the number of potential falsifiers of T' becomes greater. These potential falsifiers involve further auxiliary theories T'', and so on. But $T^i \neq T^k$ for any $i \neq k$ as circularity must be avoided. The result is that we are involved in an infinite regress unless we admit that there is some T^i without potential falsifiers. And as tests are carried out, this is not only a possibility but a fact of scientific procedure: every test involves metaphysical auxiliary assumptions.

A look at the history of science convinces one that this abstract scheme corresponds quite closely to reality. Thus some of Galileo's arguments against his opponents were based upon what was seen through the telescope. At the time in question the auxiliary theory involved, namely optics (physical and physiological), was non-existent and *a fortiori* metaphysical. Similarly, cosmological hypotheses are often measured by their agreement or non-agreement with the red shift of distant galaxies interpreted as a Doppler effect. This interpretation is again without potential falsifiers. Both abstract considerations and historical inquiry teach us that many tests involve metaphysical assumptions.

Turning the argument around, we now realize that we can increase the strength of experimental refutations by replacing these metaphysical assumptions with scientific theories, that is, by again developing alternatives to the theories under test: decisive refutation is impossible without proliferation (this is one of the features which distinguish Popper's criterion of falsifiability from such anticipations as may be found in Peirce, Dubislav, and others).

To sum up, proliferation is required both in order to strengthen our tests and in order to bring to light refuting facts that would otherwise remain inaccessible. The progress of science is unthinkable without it.

OUTLINE OF A HUMANITARIAN SCIENCE

We can now answer the question we asked before: is it possible to retain our childhood dreams, to develop them in a pleasing and fruitful fashion without losing touch with reality? The reply is clearly affirmative. Proliferation means that there is no need to suppress even the most outlandish product of the human brain, and science, far from giving comfort to the doctrinaire, will profit from such an activity and is unthinkable without it. Tenacity means that one is encouraged not just to follow one's inclinations but to develop them further, to raise them, with the help of criticism (which involves a comparison with other existing alternatives), to a higher level of articulation and thereby to raise oneself to a higher level of consciousness. The interplay between proliferation and tenacity also amounts to the continuation, on a new level, of the biological development of the species, and it may even increase the tendency for useful *biological* mutations. It may be the only possible means of preventing our species from stagnation. This I regard as the final and most important argument against the traditional theories of knowledge. Such theories are not only ill-conceived, and ill-fitted to the task of conquering nature, but their defence is also incompatible with a humanitarian outlook.

5
Experts in a free society

Jargon, jargon everywhere.
And all of it doth stink!

Let me start with a confession.

I wrote this paper in a fit of anger and self-righteousness caused by what I thought were certain disastrous developments in the sciences. The paper will therefore sound a little harsh, and it will perhaps also be a little unjust. Now while I think that self-righteousness has no positive function whatever and while I am convinced that it can only add to the fear and to the tensions that already exist, I also think that a little anger can on occasions be a good thing and can make us see our surroundings more clearly.

I think very highly of science, but I think very little of experts, although experts form about 95 per cent or more of science today. It is my belief that science was advanced, and is still being advanced by *dilettantes* and that experts are liable to bring it to a standstill. I may be entirely wrong in this belief of mine, but the only way to find out is to tell you. Therefore, with my apologies, here is my paper.

My attitude towards experts is as follows: as long as we are strict with votes as we are today when we still question giving the vote to eighteen-year-olds because of their alleged immaturity, we must certainly deny the vote to experts because of their *actual* immaturity. We must wait, until they grow up, until they become mature and responsible, that is, until they become *dilettantes*. A free society, of course, recognizes the value of *any* particular mode of life, and it therefore also recognizes the value of immaturity. A free society will therefore give the vote to eighteen-year-olds as well as to experts, but it will most carefully watch the latter since so much depends on their activity. (Besides, a free society must keep itself well informed about its hardened ingredients.) So, experts will certainly have the vote; they will certainly be listened to as every other citizen will be listened to, but they will receive none of the special powers which they would so dearly love to possess. Laymen will look after their affairs and will make the decisions which must be made if we want to apply science to society. Laymen will control science – and no harm will come of it. This is my attitude. Now let me explain it.

An expert is a man, or a woman, who has decided to achieve excellence, supreme excellence in a narrow field at the expense of a balanced development. He has decided to subject himself to standards which restrict him in many ways – his style of writing and the patterns of his speech included – and he is prepared to conduct most of his working life in accordance with these standards. He is not averse to occasionally venturing into different fields, to listen to fashionable music, to adopt fashionable ways of dressing (though the business suit still seems to be his favourite uniform, in this country and abroad), or to seduce his students. However, these activities are aberrations of his private life, they have no relation whatever to what he is doing as an expert. A love for Mozart, or for *Hair* will not, and must not, make his physics more melodious, or give it a better rhythm. Nor will an affair make his chemistry more colourful.

This separation of domains has very unfortunate consequences. Not only are special subjects voided of ingredients which make a human life beautiful and worth living, but these ingredients are impoverished, too; emotions become crude and thoughtless, just as thoughts become cold and inhumane. Indeed the 'private parts' of one's existence suffer much more than does one's official capacity. Every aspect of professionalism has its watchdogs; the slightest change, or threat of a change is examined, broadcast, warnings are issued, and the whole tremendous machinery moves at once in order to restore the status quo. Who takes care of the quality of our emotions? Who watches those parts of our language which are supposed to bring people together more closely – where one gives comfort, understanding, and perhaps a little *personal* criticism? There are no such agencies. The result is that professionalism takes over even here.

Let me give you some examples.

In 1610 Galileo reported for the first time his invention of the telescope and the observations he made with it. This was a scientific event of the first magnitude, far more important than anything we have achieved in our megalomaniac twentieth century. Not only was a new and very mysterious instrument introduced to the learned world (it was introduced to the *learned* world, for the essay was written in Latin), but this instrument was at once put to a very unusual *use*: it was directed towards the sky. And the results, the astonishing results quite definitely seemed to support the new theory which Copernicus had suggested over fifty years earlier and which was still very far from being generally accepted. How does Galileo introduce his subject? Let us hear:

> About 10 months ago a report reached my ears that a certain Dutchman had constructed a spyglass by means of which visible objects, though very distant from the eye of the observer, were distinctly seen as if nearby. Of this truly remarkable effect several experiences were related, to which some persons gave credence while others denied them. A few days later the report was confirmed to me in a letter from a noble Frenchman at Paris, Jacques

> Badovere, which caused me to apply myself wholeheartedly to inquire
> into the means by which I might arrive at the invention of a similar
> instrument ...[1]

and so on. We start with a personal story, a very charming personal story
which slowly leads us to the discoveries, and these are reported in the same
clear, concrete, colourful way: 'There is another thing', writes Galileo,
describing the face of the moon, 'which I must not omit, for I beheld it not
without a certain wonder; this is that almost in the centre of the moon
there is a cavity larger than all the rest, and perfectly round in shape. I have
observed it near both the first and last quarters, and have tried to represent
it as correctly as possible in the second of the above figures ...' and so on.

Galileo's drawing attracts the attention of Kepler who was one of the
first to read Galileo's essay. He comments: 'I cannot help wondering about
the meaning of that large circular cavity in what I usually call the left
corner of the mouth. Is it a work of a nature, or of a trained hand? Suppose
that there are living beings on the moon (following the footsteps of
Pythagoras and Plutarch I enjoyed toying with this idea, long ago ...) It
surely stands to reason that the inhabitants express the character of their
dwelling place, which has much bigger mountains and valleys than our
earth has. Consequently, being endowed with very massive bodies, they
also construct gigantic projects ...'[2] and so on.

'I have observed'; 'I have seen'; 'I have been surprised'; 'I cannot help
wondering'; 'I was delighted' – this is how one speaks to a friend or, at any
rate, to a live human being.

The awful Newton, who more than anyone else is responsible for the
plague of professionalism from which we suffer today, starts his first paper
on colours in a very similar style:

> ... [I]n the beginning of the year of 1666 ... I procured me a triangular glass
> prism, to try therewith the celebrated phenomena of colours. And in order
> thereto having darkened my chamber and made a small hole in my window
> shuts, to let in a convenient quantity of the sun's light, I placed my prism at
> his entrance, that it might be thereby refracted to the opposite wall. It was at
> first a pleasing divertissement to view the vivid and intense colours produced
> thereby; but after a while applying myself to consider them more circum-
> spectly, I became surprised to see them in an *oblong* form ... [3]

Remember that all these reports are about cold, objective, 'inhuman',
inanimate nature; they are about stars, prisms, lenses, the moon; and yet they

---

[1] [Quoted from 'The Starry Messenger' in S. Drake (ed.), *Discoveries and Opinions of Galileo* (New
York: Doubleday Anchor, 1957), pp. 28–9. The quotation in the paragraph which follows is
from the same text, p. 36. (Ed.)]

[2] [Quoted from *Kepler's Conversation with Galileo's Sidereal Messenger*, trans. E. Rosen (New York,
1965), p. 28. (Ed.)]

[3] [Quoted from *Newton's Philosophy of Nature: Selections from his Writings* (New York: Hafner,
1953), p. 68. (Ed.)]

are described in a most lively and fascinating manner, communicating to the reader an interest and an excitement which the discoverer felt when first venturing into strange new worlds.

Now compare with this the introduction to a recent book – a best-seller even – *Human Sexual Response* by Masters and Johnson. I have chosen the book for two reasons. First, because it is of general interest. It removes prejudices which influence not only the members of some profession, but the everyday behaviour of a good many apparently 'normal' people. Secondly, because it deals with a subject that is new and without special terminology. Also, it is about men rather than about stones or prisms. So, one would expect a beginning even more lively and interesting than that of Galileo, Kepler, or Newton. What do we read instead? Listen: 'In view of the pervicacious gonadal urge in human beings, it is not a little curious that science develops its sole timidity about the pivotal point of the physiology of sex. Perhaps this avoidance ...'[4] and so on. This is no longer human speech. This is the language of the expert.

Note that the subject has completely left the picture. Not: '*I* was very surprised to find' or, as there are two authors, '*We* were very surprised to find,' but '*It is* surprising to find ...', only, not expressed in these simple terms. Note also to what extent irrelevant technical terms intrude and fill the sentences with antediluvian barks, grunts, squeaks, belches. A wall is erected between the writer and his readers not because of some lack of knowledge, not because one does not know who the reader is, but in order to make utterances conform to some curious professional ideal of objectivity. And this ugly and inarticulate idiom turns up everywhere and takes over the function of the most simple and straightforward description.

Thus on page 65 of the book we hear that the female, being capable of multiple orgasm, must often masturbate after her partner has withdrawn in order to complete the physiological process that is characteristic for her. And, so the authors want to say, she will stop only when she gets tired. This is what they *want* to say. What they actually say is: 'usually physical exhaustion alone terminates such an active masturbatory session'. On the next page the male is advised to *ask* the female what she wants or does not want rather than try to guess it on his own. 'He should ask her' – this is what our authors want to convey. What is the sentence that actually lies there in the book? Listen: 'The male will be infinitely more effective if he encourages vocalisation on her part.' 'Encourages vocalisation' instead of 'asks her'. Well, one might want to say, the authors want to be precise, and

---

[4] [W. H. Masters and V. E. Johnson, *Human Sexual Response* (Boston: Little Brown and Co., 1966), p. v. Although Feyerabend is quoting the very first words of the book's preface, he does not make it clear that these words are themselves being quoted, by Masters and Johnson, from R. L. Dickinson and H. H. Pierson, 'The Average Sex Life of American Women', *Journal of the American Medical Association*, vol. 85, 1925. (Ed.)]

they want to address their fellow professionals rather than the general public and, naturally, they have to use a special lingo in order to make themselves understood. Now as regards the first point, precision, remember that they also say that the male will be *'infinitely* more effective'. Which, considering circumstances, is not a very adequate statement of the facts. And as regards the second point we must say that we are not dealing with the structure of organs, or with special physiological processes which might have a special name in medicine, but with an ordinary affair such as *asking*. Besides, Galileo and Newton could do without a special lingo although the physics of their time was highly specialized and contained many technical terms. They could do without a special lingo because they wanted to start afresh. Masters and Johnson find themselves in the same position, but they cannot speak straight any more, their linguistic talents and sensibilities have been distorted to such an extent that one asks oneself whether they will ever be able to write normal English again.

The answer to this question is contained in a little pamphlet which I have saved and which contains the report of an ad hoc committee formed for the purpose of examining rumours of police brutality during some rather restless weeks in Berkeley (winter 1968/9). The members of the committee were all people of good will. They were interested not only in the academic quality of life on campus, they were even more interested in bringing about an atmosphere of understanding and of compassion. Most of them came from sociology or from related fields; that is, they came from fields which deal not with lenses, stones, stars, as did Galileo in his beautiful little book, but with humans. There was a mathematician among them who had devoted considerable time to setting up and defending student-run courses and who finally gave up in disgust – he could not change the 'established academic procedures'. How do these nice and decent people write? How do they address those to whose cause they have devoted their spare time and whose lives they want to improve? Are they able to overcome the boundaries of professionalism at least on this occasion? Are they able to *speak?* They are not.

The authors want to say that policemen often make arrests in circumstances when people are bound to get angry. They say: 'When *arousal* of those present is the inevitable consequence'. 'Arousal'; 'inevitable consequence' – this is the lingo of the laboratory; this is the language of people who habitually mistreat rats, mice, dogs, rabbits and carefully notice the effects of their mistreatment, but the language is now applied to humans, too; to humans, moreover, with whom one sympathizes, or says one sympathizes, and whose aims one supports. They want to say that policemen and strikers hardly talk to each other. They say: 'Communication between strikers and policemen is non-existent.' Not the strikers, not the police, not people are the centre of attention, but an abstract process, 'communication', about which one has learned a thing or two and with which one feels more at ease than with living human beings. They want to say that more than eighty people participated in the venture, and that the report contains the common elements of what about thirty of them have written. They say: 'This report tries to reflect a consensus from the thirty reports submitted by the eighty plus faculty observers who participated.' Need I continue? Or is it not already clear that the effects of professionalism are much deeper and much more vicious than one would expect at first sight? That some professionals have even lost the ability to *speak*, that they have returned to a state of mind more primitive than that of an eighteen-year-old who is still able to adapt his language to the situation in which he finds himself, talking the lingo of physics in his physics class and quite a different language with his friends in the street?

Many colleagues who agree with my general criticism of science find this emphasis upon language farfetched and exaggerated. Language, they say, is an instrument of thought which does not influence it to the extent I surmise. This is true as long as a person has different languages at his disposal and as long as he still has the ability to switch from one to another as the situation demands. But this is not the case here. Here a single and rather impoverished idiom takes over all functions and is used under all circumstances. Does one want to insist that the thought that hides behind this ugly and inhuman exterior (emphasis upon abstract processes such as 'communication' instead of living people) has remained nimble and humane?

That being an expert is a predicament and not a matter of pride was realized, long ago, by Aristotle. A free man, according to Aristotle, is a man who has a sense of balance. He has a sense of perspective. He is well informed, in politics, in the sciences, in the arts. He gives some weight to all these things, he lets all of them influence his being to some extent. Men think – but they are also capable of emotions. They have <an> interest in politics – but they also wonder about the stars. They want power – but they also want on occasions to submit to a higher authority. None of these interests, none of these subjects can demand exclusive attention, and each of them must be pursued with restraint. This restraint cannot be achieved abstractly, by devoting oneself to one subject and thinking that there may be a limit to it. Such thinking will soon lose its effectiveness and will become an empty formula unless it is supported by the concrete experience of what goes on outside the limit. It is this concrete experience which prevents a man from becoming narrow-minded and partial in the sense of being part of a man only; it is this concrete experience which prevents him from becoming a slave. In other words: you can be a free man, you can achieve and yet retain the dignity, the appearance, the speech of a free man only if you are a *dilettante.* 'Any occupation, art, science', writes Aristotle, 'which makes the body, or soul, or mind ... less fit for the practice or exercise of virtue, is vulgar; therefore we call those arts vulgar which tend to deform the body, and likewise all paid employments, for they absorb and degrade the mind. There are some liberal arts quite proper for a free man to acquire,' Aristotle continues, '*but only to a certain degree,* and if he attends to them too closely, in order to attain perfection in them, the same evil effect will follow' – he will become a slave in mind, and soon in actual position as well. Just remember to what extent the academic profession makes slaves of its members, especially of the untenured ones, and also remember how greedy and intolerant those slaves become once they get a whiff of freedom, or what they think is freedom, viz., tenure. 'The organisation of science', writes Robert Merton on precisely this point,

> operates as a system of institutionalized vigilance, involving competitive co-operation. It affords both commitment and reward for finding where others have erred or have stopped before tracking down the implications of their results or have passed over in their work what is there to be seen by the fresh eye of another. In such a system, scientists are at the ready to pick apart and appraise each new claim to knowledge. This unending exchange of critical judgement [which can become quite nasty – *vide The Double Helix,* or the reaction to Velikovsky], of praise and punishment, is developed in science to a degree that makes the monitoring of children's behaviour by their parents seem little more than child's play.[5]

There are, of course, wandering minstrels who try to bewitch the onlookers by praising the beauties of science, the joys of discovery, the essentially human character of the search for knowledge and truth – or whatever other titles they choose for their arias. I am afraid they are singing about a time that has long gone by, and their songs are not melodious enough to let us forget the present squalor.

To sum up: experts today are excellent, useful, irreplaceable, but mostly nasty, competitive, ungenerous slaves, slaves both in mentality, speech, and in social position.

Now, what I have said so far is only one side of the matter and, although quite depressing, it is the most innocuous one. It is quite depressing to see with what fervour thousands of young people throw themselves into special subjects where they are trained and trained and trained by receiving now punishment, now a pat on the head until they are hardly distinguishable from the computers whom they want to approach in efficiency – except that being human they have to add self-righteousness, lack of perspective, Puritanism, and atrocious professional jokes to what they are pleased to call

[5] R. K. Merton, 'Behavior Patterns of Scientists', reprinted in his *The Sociology of Science* (Chicago: University of Chicago Press, 1973), p. 339.

the various steps of their reasoning. Now, the peculiar situation in which we find ourselves today is that these inarticulate and slavish minds have convinced almost everyone that they have the knowledge and the insight not only to run their own playpens, but large parts of society as well, that they should be allowed to educate children, and that they should be given the power of doing so without any outside control and without supervision by interested laymen. One of the most basic elements of the scientific ideology (and of expert ideology in general) is that only a scientist can understand what a scientist is up to and that only a scientist can decide how a scientist should be employed. For example, only a scientist can know how his subject should be taught and only he knows how important it is when compared with other subjects. It is this demand of the experts which I want to examine in the rest of my paper. Should we allow a bunch of narrow-minded and conceited slaves to tell free men how to run their society? What arguments do they have to demand such meekness from us? What arguments do they have to demand not only that their own particular business should be exempt from inspection by non-experts (though, of course, it should be *financed* by them), but also that their religion should become a state religion and that the education of the young should be left in their hands entirely? What arguments do they have to support their impudent demand that evolution should replace Genesis as a view of man and why are there theologians who try to redefine their subject so that no clash with science can ever arise? Has it been *proved* that scientific theories are more correct than anything that follows from a literary interpretation of the Bible? What are the proofs? Let us see!

One decisive step in the development of science was the so-called scientific revolution of the sixteenth and seventeenth centuries. There are still many people who believe that this event was the result of a radical empiricism and who think that it occurred only because one man decided to eliminate views not in agreement with observation and reasonable generalizations therefrom. Hardly anyone is prepared to admit that Copernicus might have been in greater trouble than the Aristotelian–Ptolemaic cosmology and that his views might have succeeded because of ad hoc changes in the evidence, specious arguments, and a lot of hot air. Galileo's strong faith in Copernicus, his ebullience, his propagandistic ability, his willingness to cheat played a decisive role in the battle that was about to begin. It is interesting to see how suspicious Galileo is of *experience* and how often he prefers an interesting and intriguing *hunch* to a clear and straightforward observation. His suspicion has various sources. It is connected with the fact that experience played a role in the magical tradition which he despised. Agrippa, Trithemius, the legendary Faust – they all point out that reason has its limits and that it must on occasion be supplemented by a mysterious, magical, but still quite trustworthy source, viz.: experience. 'Formal forces are called occult forces', writes Agrippa in his *De Occulta Philosophia*, 'for their causes are hidden from us; human reason cannot examine them in all directions and this is why philosophers have learned them from experience, not from deep thinking. . .' Alchemy<,> which deals with occult forces, is firmly empirical. And so is the art of discovering witches. Asked whether his ability to find witches under the most difficult and trying circumstances 'proceeded from profound learning, or from much reading of learned authors', Master Matthew Hopkins, most excellent, most wise, and most terrible witchfinder General of the forties (the *sixteen*-forties) replies: 'From neither or both, *but from experience* which though it is meanly esteemed of, is yet the surest and safest way to judge by' (*The Discoverie of Witches*, Answer Three). Here we are already approaching Bacon whose empiricism has much in common with the magical tradition but who is also influenced by the Lutherans and by their search for a firm foundation of the faith. This is easily seen from passages such as the following:

> We have now treated of each kind of idols and of their qualities, all of which must be abjured and renounced with firm and solemn resolution, and the understanding must be completely freed and cleared of them, so that the access to the kingdom of man, which is founded on the sciences, may resemble that <to> the kingdom of heaven, where no admission is conceded except to children. (*Novum Organum*, §68)

Such magical and enthusiastic types of empiricism which exclude thought from a large area of knowledge are viewed with considerable suspicion and distaste by Galileo, both in his early works on motion, and later on (remember that he rejects the moon-theory of the tides because of its astrological flavour, despite the existing evidence). He even rejects the sober empiricism of Aristotle, the only empirical philosophy, incidentally, that has ever been developed in a rational way. Aristotle explains *what* experience is, and he gives reasons *why* it should be regarded as a foundation: experience, according to Aristotle, is what we perceive under normal circumstances, with our senses in good order, and what we then describe in an idiom that is familiar to all. It is trustworthy because man and the universe are attuned to each other, because they are in harmony. The harmony is not denied by Galileo, but he doubts that it can help us to discover the basic laws of the world we live in. The phenomena which we perceive depend on those universal laws, but they also depend on the special conditions which make them appear. For example, our perception of the stars depends on the properties of light, on the conditions of the terrestrial atmosphere, as well as on the idiosyncrasies of our senses. Similarly, our perception of celestial motion depends on the actual motion of the stars, on the special conditions of our observation platform, the earth, and again on the idiosyncrasies of the senses. It is therefore necessary to *analyse* the phenomena, to subdivide them, and to subtract from them what is due to the special conditions of their origin.

This analysis is carried out by Galileo with great theoretical skill, and its results are presented and defended with even greater forensic talent. Galileo himself soon takes the Copernican cosmology for granted. Hence, one boundary condition of the analysis which he sets himself and which influences his research in dynamics and in optics is that it must lead to the Copernican universe. It is this boundary condition and not any profound and complicated experimental work which is responsible for the gradual change of his dynamics from an interesting type of impetus theory into a wholly new account where motion, and even the motion of a large and sluggish piece of matter like the earth, can occur without any moving force whatever. The boundary condition also leads to a redefinition of dynamic terms with the consequence that observation now ceases to conflict with Copernicus. *All these changes are purely ad hoc.* Moreover, they break the very close connection between observation and theory that was characteristic of the Aristotelian philosophy; observation and theory drift apart, leaving a sizeable chasm between them. The chasm is noticed and filled partly by the promise of further research, partly by spurious experiments, partly by an appeal to what 'the reader surely knows but has forgotten' (a phrase that occurs rather frequently in the *Dialogue*), partly by reference to new, surprising phenomena which, though puzzling in themselves and without any theoretical explanation yet seem to fit perfectly into some of the vacant places. And now Galileo reverses the whole procedure, he starts from facts, plausible conjectures, adds further facts, appeals to the reader's common-sense until the Copernican doctrine arises as an almost inescapable conclusion. This is a fascinating performance to watch, for it shows that science at its best demands all talents of man, his critical sense, as well as his literal ability, his prejudices as well as his caution, his arguments as well as his rhetoric, his honesty as well as his will to deceive, his mathematical ability as well as his artistic sense, his modesty as well as his greed, – it shows that science at its best demands all these talents and ennobles them by making them an essential part of the movement toward a better understanding of our material and intellectual condition. Today we can give some reasons why such an opportunism has always a chance to succeed. A cosmology and the existing evidence may be out of phase in the sense that the evidence depends on ancient views while the cosmology is a step forward. In this case the cosmology will be in trouble, not because it does not represent the truth, but because the customary measure of the truth – the evidence – is contaminated. This being a possibility it is quite legitimate to divert attention from the evidence, to make propaganda for an apparently refuted view, to reinterpret the evidence in its light, and to transfer the general enthusiasm for observations to the changed evidence.

This is the heroic time of science when one can be both a scientist and a man in the full sense of the word, when a pleasant and melodious style, full of personal allusions and entertaining asides is not yet regarded as a hindrance to clear thought, and when the best scientist is at the same time the best and the most outstanding *dilettante*. Expert knowledge *does* exist but it is not produced by people who have devoted themselves to a narrow field for their whole life, to the exclusion of everything else, but by people who have studied a subject for a year or two, who have a sense of perspective, and who can therefore give a well-rounded account of special fields also. (Such people exist even today, although their number is vanishingly small. Thus<,> describing the origin of the various textbooks of which he was author or co-author, Max Born points out 'that in order to write a learned volume one need not specialise in the subject but only grasp the essentials and do some hard work'. He continues: 'I never liked being a specialist and have always remained a dilettante, even in what were considered my own subjects. I would not fit into the ways of science today [the 1960s], done by teams of specialists. The philosophical background of science has always interested me more than its special results . . .').[6]

Now after all this splendour – where does the present squalor come from? There are many sources, most of them still unexamined. In what follows I shall discuss only one.

One particular element of the expert ideology that existed at all times and that plays a role in such different traditions as the hermetic tradition and the empiricism of the nineteenth and the twentieth centuries is the belief that success and progress can be achieved with the help of *special methods* only. Simon Magnus, Galileo, Newton – they all insinuate that there are special ways of obtaining knowledge, and that they have succeeded by using these special ways. Experience may be emphasized, and it is emphasized by the hermetic tradition (see the quotation from Agrippa above) and by the altogether different tradition of critical rationalism. It would be extremely interesting to examine this belief further in a method, and to inquire into its origin. Only a little research has been done, and the results that have been found have often been distorted in one way or another. However, our interest here is not in the *origin* of the belief; what interests us is its *effect* on the development of the sciences. And this effect can be easily ascertained.

Neither Galileo, nor Kepler, nor Newton use specific and well-defined methods. They are eclectics, methodological opportunists. Of course, each individual has what one could call a special *style* of research that gives his efforts some kind of unity, but the style changes from one individual to the

---

[6] M. Born, *My Life and Views* (New York: Scribner's, 1968), p. 22 (ed.).

next, and from one piece of research to another. Galileo on occasions acts like an empiricist while on other occasions he seems to be a tough-minded rationalist with no concern for observational results. Newton proceeds very differently in his research on celestial mechanics and in his research on optics. Compare Newton with Hooke, and you will see the variety of attitudes and of styles that existed in the Royal Society toward the second third of the seventeenth century. So, looking at the actual historical situation we see that science was advanced in many different ways and that scientific problems were attacked by many different methods. In practice the only principle that is constantly adhered to seems to be: *anything goes.*

Nor is it difficult to understand why this should be so. A scientist finds himself in a complex historical situation. There are observations, attitudes, instruments, ideologies, prejudices, errors, and he is supposed to improve theories and change minds under the highly individualized circumstances created by the interplay of all these factors. Instruments as well as people must be coaxed into giving the proper response, taking into consideration that no two individuals (no two scientists; no two pieces of apparatus; no two situations) are ever exactly alike and that procedures should therefore be allowed to vary also. In many ways a good scientist has to be like a politician who possesses an intuitive grasp of the objective situation and of the mood of his audience, and who has to make the best of both if he wants to get his views across. Or he is like a prize-fighter who tries to discover the idiosyncrasies, the weaknesses, the advantages, the special moves of his opponent in order to be able to adapt his style to them. Considering the complexity of the world he lives in, this eclecticism of the scientist, this 'ruthless opportunism' (Einstein), is not just an expression of human inconstancy and folly. It is the only type of behaviour that has a chance of succeeding.

Now it is interesting to see how great scientists, while intuitively adopting a methodological opportunism, or anarchism of this kind almost always act as if they had followed a specific and well-defined method. We have already described the case of Galileo. He changes ideas, bends concepts, reinterprets laws and observations to fit the Copernican view; he uses ad hoc hypotheses, but he also tries to create the impression that he has arrived at this view in a systematic manner, relying now on mathematics, now on observation, now on simple and straightforward commonsense. The case of Newton is even clearer, for he spells out the methodology that has allegedly guided him in his research. There are three different levels: phenomena, laws, hypotheses. These levels are distinct, and must be kept distinct. Hypotheses must never interfere with the phenomenal level, nor must they be used either to suggest, or to constitute laws. Laws are derived from phenomena, and are explained with the help of hypotheses. All this is familiar stuff, especially for those who have read Nagel's *Structure of Science*

(London: Routledge, 1961). But Newton not only preaches methodology, he also presents his results in a form that perfectly fits the pattern of research he recommends. He thereby convinces everyone that the way from phenomena to laws to hypotheses is indeed the only way for a scientist to follow. Every scientist now either tries to proceed in this manner, he tries to find laws by collecting phenomena and looking for suitable derivations, or he tries at least to tell the story of his discoveries in this way, no matter how irrational and whimsical the actual sequence of events. There arises then a period of schizophrenia when a scientist does one thing and says, and believes, that he does another.[7] Not everyone can live such a double life, and many people just carry out one experiment after another and hope for the best. Some of them make valuable contributions not because they have found the one and only correct method after all, but because any method, as we have said (including the silly method of multiplying experiments) has a chance of producing results. As science progresses and becomes more complex, it also becomes more and more difficult to fit it into the simple Newtonian pattern. The pattern gradually dissolves and is replaced by increasingly vague and ritualistic statements. For example, the notion of a phenomenon and the more general notion of experience is widened so that in the end it can contain almost any law and almost any hypothesis. One realizes that the scientific method is more complicated than one had thought, and that it cannot be captured by a simple set of rules. Yet, despite all these difficulties one still believes that there is something like a method, but one now assumes that it is hidden in the ongoing process of science, and that it can be absorbed by immersing oneself in the process, and by participating in it in a spirit of complete and faithful conformism. It is this myth of a hidden method rather than any sound evaluation of the nature of science that underlies the expert's demand for special powers, that supports his claim that scientific knowledge has sufficient authority to resist and eliminate any extrascientific ideas.

---

[7] For historical examples <see> the following essays of mine: 'Classical Empiricism' in R. E. Butts (ed.) *The Methodological Heritage of Newton* (Oxford: Blackwell, 1970); 'Problems of Empiricism' in R. G. Colodny (ed.) *Beyond the Edge of Certainty* (New Jersey: Prentice-Hall, 1965); 'Problems of Empiricism, Part II' in R. G. Colodny (ed.) *The Nature and Function of Scientific Theories* (Pittsburgh: Pittsburgh University Press, 1970); 'Against Method' in M. Radner and S. Winokur (eds.), *Minnesota Studies in the Philosophy of Science, volume IV* (Minneapolis: University of Minnesota Press, 1970). These essays deal mainly with Newton and with Galileo. Kuhn and his collaborators have analysed more recent episodes in the history of science and have made some surprising discoveries. <See> the report on the interviews on pp. 3ff of T. S. Kuhn, J. L. Heilbron, P. Forman and L. Allen, *Sources for the History of Quantum Physics. An Inventory and Report* (Philadelphia: American Philosophical Society, 1967), as well as Paul Forman, 'The Discovery of the Diffraction of X-Rays by Crystals: A Critique of the Myths', *Archive for the History of the Exact Sciences*, vol. 6, 1969, pp. 38–71, and John L. Heilbron and Thomas S. Kuhn, 'The Genesis of the Bohr Atom', *Historical Studies in the Physical Sciences*, vol. 1, 1969, pp. 211–90.

Now this myth could prevail only as long as science seemed to be perfect and free from error, as long as there were only minor disturbances, minor corrections, but no major breakdowns. For such disturbances could always be ascribed to inattention or use of the improper method, and besides, they were soon forgotten and eliminated from the official histories of the subject. These histories were, and still are success stories, reports of an uninterrupted flood of discoveries and additions to a solid bulk of knowledge that is occasionally subjected to minor corrections, but which is essentially sound and invulnerable.

The situation changes drastically with the scientific revolution of the twentieth century, with the arrival of the quantum theory and the theory of relativity. For it now turns out that a great and successful scientific world view can be entirely wrong and that it might be necessary to replace not just a constant, or a peripheral law, but fundamental concepts one had used for describing the most common and the most easily observable events.

Turning back to history with this new insight one realizes that the official success story was but the result of wishful thinking, that science has always progressed through catastrophes and intellectual upheavals, and that not a single scientific theory is free from serious trouble. *There is no method, and there is no authority*. Of course, there still remains an almost religious faith in the excellence of science and in the supremacy of its results. But it is clear that a free society will have to treat this faith like all other beliefs, such as astrology, or black magic; it will guarantee its adherents freedom of expression, but it will not grant them any of the special powers they would so dearly love to possess.

But is it not utterly foolish to adopt such an attitude? Is it not clear that science has produced innumerable valuable results while astrology has produced nothing? Is it not safer in the case of serious illness to trust the judgement of a physician rather than the judgement of a witch, and should physicians not therefore receive a special position in our society? Was it black magic or physics that brought man to the moon and, if the former, has not therefore physics proved its case and acquired the right for special treatment? These are the questions which are thrown at the impudent critic who dares to suggest that a scientist is but a citizen and that whatever special rights he obtains must be based on the judgement of other citizens, laymen included. It is not difficult to reply to these questions.

To start with, it is not suggested that a hospital should employ qualified physicians side by side with witches, or that the space programme should consult experts in levitation as well as physicists and astronomers, giving them the same authority. What *is* denied is that a judgement of this kind should be left in the hands of the experts entirely, and that laymen, or experts of a different kind (experts in black magic, for example) should not have any say in the matter. A hospital serves a community, and so it must be left to the members of this community, experts and laymen alike, to come to a decision. Besides, experts are a prejudiced party in the dispute, they want to get jobs which are respectable and well paid, and so quite naturally they will praise themselves and condemn others. Their ideas must therefore be balanced by the ideas of outside observers. Will these outside observers have sufficient insight into a complex situation to be able to come to a useful conclusion? The reply is that scientists don't have such insight either, they often disagree on fundamental matters, and periods of agreement may be due to a large dose of conformism rather than to a shared truth. It is quite customary to let a patient, or the relatives of a patient make a decision about a complex operation not because they have special knowledge, but because they are the concerned party and because the responsibility should be left to them. On the larger scale the situation is exactly the same. Only, it is not just a matter of responsibility that is at stake here. Special professions often have a special and quite narrow vision of their subject matter. Physicians deal only with certain aspects of man, they concentrate on the physical side and so it is quite possible that they lack knowledge that has been assembled by some non-professionals. Paracelsus learned from witches, Galileo learned from gunners and carpenters, Edison and the Wright brothers succeeded despite the opposition of contemporary science and nobody can say that this process of learning has come to an end or that there are no further discoveries to be made outside a certain profession. Also, the need to communicate with laymen, to explain to them their particular business and the reasons for the convictions they hold will force the experts to speak more simply; it will force them to relearn a language which they have almost forgotten and which they have replaced by an ugly and narrow-minded idiom. It will make their language more humane; it will make them more humane. All these are desirable developments – but they will occur only if we abandon the great and unreasonable reverence and virtual fear we have of experts and adopt instead the more sensible view that experts are humans just as we are; that they have therefore the ability to produce bright ideas and the related ability to commit grievous mistakes.

6

Philosophy of science: a subject with a great past

(**1**) While it should be possible, in a free society, to introduce, to expound, to make propaganda for any subject, however absurd and however immoral, to publish books and articles, to give lectures on any topic, it must also be possible to *examine* what is being expounded by reference, not to the *internal* standards of the subject (which may be but the method according to which a particular madness is being pursued), but to standards which have the advantage of being simple, commonsensical, and accepted by all. Using such standards as a basis of judgement we must confess that much of contemporary philosophy of science and especially those ideas which have now replaced the older *epistemologies* are castles in the air, unreal dreams which have but the name in common with the activity they try to represent, that they have been erected in a spirit of *conformism* rather than with the intention of influencing the development of science, and that they have lost any chance of making a contribution to our knowledge of the world. (The medieval problem of the number of angels at the point of a pin had some rather interesting ramifications in optics and in psychology. The problem of 'grue' has ramifications only in the theses of those unfortunate students who happen to have an engruesiast for a teacher). This is my opinion. Let me now give some reasons for it.

(**2**) The scientific revolution of the sixteenth and seventeenth centuries is characterized, among other things, by a close collaboration between science and philosophy. This is a direct consequence of the way in which science was debated both in antiquity and in the Middle Ages. The reaction against 'medieval science', which in many cases was but a reaction against certain petrified aspects of it, leads to the development of new philosophical principles. It does not lead to a split between science and philosophy. The new philosophy that is being gradually developed is of course used to expose and to remove the hardened dogma of the schools. However, it has also a quite decisive role in building the new science and in defending new theories against their well-entrenched predecessors. For example, this philosophy plays a most important part in the arguments about the Copernican system, in the development of optics, and in the construction of a new and non-Aristotelian dynamics. Almost every work of Galileo is a mixture of philosophical, mathematical, and physical principles which collaborate intimately without giving the impression of incoherence. This is the *heroic time* of the scientific philosophy. The new philosophy is not content just to *mirror* a science that develops independently of it; nor is it so distant as to deal just with alternative *philosophies*. It plays an essential role in building up the new science that was to replace the earlier doctrines.[1]

(**3**) Now it is interesting to see how this active and critical philosophy is gradually replaced by a more conservative creed, how the new creed generates technical problems of its own which are in no way related to specific scientific problems (Hume), and how there arises a special subject that codifies science without acting back on it (Kant). One can say, without too much simplification, that the change is essentially due to *Newton*. Newton invents new theories, he proposes a radically empiricist methodology, and he claims that he has obtained the former with the help of the latter. He supports his claim by a manner of presentation that seems indeed to suggest, at least at first sight, that his optics and his celestial mechanics are the perfect results of a perfect method perfectly applied. Having convinced most of his contemporaries, he creates additional support both for his science (it has been obtained in a methodologically sound way and must therefore be free from major mistakes) and for his methodology (it has led to perfect scientific results and must therefore be the correct method).[2] Of course, his presentation is quite misleading, it is full of holes, fallacies, contradictions, and he himself violates *every single rule* he proposes. Yet it was influential enough to have blinded scientists, historians (including some very recent students of the history of optics, such as Westfall), and philosophers alike.[3] 'Experience', from now on, means either the results of Newton's experiments as described by him (optics), or the premises of his deductions (celestial mechanics), but it *also* means, by virtue of Newton's connecting manoeuvre, the safe, irrevocable and gradually expanding basis of scientific reasoning. Small wonder that thinkers who seemed to sense a flaw but who lacked either the patience or the talent to combine their critical intuition with spectacular scientific discoveries were not heard, and were increasingly isolated.[4] To survive, they changed their target from science to philosophy and thus created (or, rather, continued – for there

---

[1] For details concerning Galileo and his difference from Descartes and Bacon, see my essay 'Bemerkungen zur Geschichte and Systematik des Empirismus' in Paul Weingartner (ed.), *Grundfragen der Wissenschaften und ihre Wurzeln in der Metaphysik* (Salzburg: Anton Pustet, 1966).

[2] The development in the quantum theory from 1927 to about 1955 was of exactly the same kind.

[3] For optics see Goethe, *Theory of Colours*, which contains a very perceptive account of the ideological development just mentioned; V. Ronchi, *Histoire de la lumière* (Paris: Colin, 1956); A. I. Sabra, *Theories of Light from Descartes to Newton* (London: Oldbourne, 1967); as well as my discussion of Newton in 'Classical Empiricism' in R. E. Butts and J. W. Davis (eds.), *The Methodological Heritage of Newton* (Oxford: Blackwell, 1970).

[4] An exception was Faraday, but his background philosophy remained almost completely unknown.

was always a tradition that developed philosophy out of its own problems and with only the most tenuous relation to science) a self-sufficient subject, content with discussing its own problems. Science, on the other hand, being separated from philosophy, had to rely on intuitions of a different and much more narrow kind. The possibility of a fundamental criticism became more and more remote. In this respect, too, the situation was surprisingly similar to the situation that exists in certain parts of science today.[5] However, there is one difference. The nineteenth century produced one philosopher who was not prepared to accept the status quo, who was not content to criticize science from the safe distance of a special subject either, but who proceeded to suggest concrete means for its change. The nineteenth century produced *Ernst Mach*.

(**4**) Ernst Mach's 'philosophy'[6] contains a general criticism of the science of his time, including the house philosophy of the contemporary Newtonians, and a philosophy of science that completely abandons the idea of a foundation of knowledge. The criticism and the positive views are illustrated by his work in the history of science where factual and epistemological considerations are once more merged in perfect harmony, and they are given strength by the exhibition of shortcomings right in the centre of the most advanced theories of the nineteenth century. The criticism and Mach's own positive suggestions have been extremely fruitful, both in the sciences and in philosophy. In science Mach's suggestions have contributed to the development of the general theory of relativity and they play an essential role in more recent discussions, now that interest in general relativity has been revived. They had also a decisive influence, not always beneficial, on the founders of the quantum theory (even Schrödinger was often heard to say, quite emphatically: 'Aber wir koennen doch nicht hinter Mach zurueckgehen!'). In philosophy it was a different story, as will be seen below.

In order to understand Mach we must distinguish most carefully (and more carefully than Mach himself did on various occasions) between his *general methodology* (s.v.v.) and the more *specific hypotheses* he used as starting points of research.

A general methodology is independent of any particular assertion about the world, however trivial, and however obvious. It is supposed to provide

---

[5] The similarity becomes even greater in view of the fact that the famous chocolate-layer-cake model of scientific knowledge that has been developed by Nagel, Hempel, and others is nothing but a more sophisticated (and less clear) repetition of Newton's views, down to the last mistake. <See> 'Classical Empiricism', footnote 8, as well as my review of Nagel in the *British Journal for the Philosophy of Science*, vol. 17, 1966, pp. 237–49.

[6] 'Above all, there is no Machian philosophy; there is, at the most, a methodology of science and a psychology of knowledge and like all scientific theories these two things must be regarded as preliminary and incomplete attempts', *Erkenntnis und Irrtum* (Leipzig: Johann Ambrosius Barth, 1905), p. vii, footnote (against Hoenigswald).

a point of view from which *all* such assertions can be judged and examined. It will not assume a dichotomy between an objective world and a perceiving subject who explores the world (using his mind and his senses) and gradually increases his knowledge of it. Such a dichotomy is presupposed by almost all science, it is the instinctual basis of everyday behaviour (at least in Western societies), and it has been professed with almost religious fervour by thinkers who otherwise pride themselves on having made criticism a principle of science and of philosophy. Yet – is it not possible that this view is mistaken? Is it not possible that it neglects or misrepresents phenomena of an intermediate nature which show that the boundary is rather ill-defined and perhaps altogether non-existent? And if we *admit* this possibility, must we not ask ourselves how such a failure of realism can possibly be detected without relying on realism in the process? At this point recourse is usually made to *sensations*, and this is quite appropriate if the existence of sensations, bundles of sensations, lawful connections between sensations, is regarded as an *alternative hypothesis* rather than an eternal measuring stick of any subject that is not explicitly about sensations. For just as the existence of the real world is a topic for critical discussion, in the very same manner the existence of sensations is a topic for critical discussion. General methodology, therefore, must refer neither to the one nor to the other (although it may provide rules for playing them off against each other) and, indeed, *no such reference is made by Mach*. According to Mach the task of science is to find simple and regular connections between *elements*. Let us analyse the various parts of this assertion.

(**5**) All regularity, says Mach, is imposed, or constructed, yet it never fits all relevant cases. This assertion is not entirely without content. It is assumed that there exists a domain where regularities are produced and another domain which has idiosyncrasies of its own and which can never be fully comprehended, or tamed, by the imposed laws. Every rule, every law, even the most precise formulation dealing with events carefully prepared is bound to have exceptions, and even a perception which at first sight seems perfectly symmetrical loses this symmetry on closer inspection. This idea of the two domains comes forth very clearly in passages such as 'What is constant, the rule, a point of departure does not exist except in our thinking', which have a decidedly Kantian flavour.[7] 'It is we who glue things together, not nature.'[8] This seems to push us irrevocably toward a theory of sensations *but it does not*, for the elements which are related in this more or less regular manner are carefully distinguished from sensations. Speaking of sensations, Mach says quite explicitly, *already entails acceptance of*

---

[7] Notebook III, February 1882, p. 82, quoted from Hugo Dingler, *Die Grundgedanken der Machschen Philosophie* (Leipzig: Johann Ambrosius Barth, 1924).
[8] Notebook I, May 1880, p. 58: 'Nur wir kleben zusammen, die Natur nicht'.

a one-sided theory.[9] Elements are not sensations. They are not perceptions either, for perceptions are rather complex entities, containing memories and attitudes as well as 'natural habits'[10] of the human species. They are certainly not material objects. *They are open places, to be filled and refilled by the results of research.* All that is asserted is therefore that complexes consisting of elements which may in turn be complexes (though perhaps complexes of a simpler kind, at least at a particular stage of knowledge) are assembled into higher units whose stability is always in question and may be upset either by new theoretical procedures, or by a change of our ordering habits, or by the realization that essential things have been left out, or by a change in the elements, and so on. Now if we want to remove as many particular assumptions as possible, if we want to arrive at truly general methodology, then we must abandon this opposition between an ordering mind and an ordered material also and must restrict ourselves to stating a *development* of prima facie simple elements which arrange and rearrange themselves, dissolve and recombine in different patterns, a view that has great similarity with what is explained in Hegel's *Logik* (except that it does not contain the more specific hypothesis of modified preservation).[11]

Now research proceeds by first filling this general scheme with more specific content and then developing consequences, always remaining critical of the particular hypotheses used. The hypothesis to which Mach appeals rather frequently identifies the elements with *sensations*. This hypothesis plays an important part in Mach's criticism of contemporary physics and philosophy. *But it is never regarded as being above criticism.* It functions rather like, let us say, the principle of Lorentz invariance<,> which is constantly used for the criticism of theories without being itself exempt from criticism.

That sensations cannot be an absolute basis for Mach becomes clear from the very title of his book *The Analysis of Sensations*. Sensations are to be *analysed*. The complexity that hides behind the simple appearance is to be discovered and reduced to other and perhaps as yet unknown elements. These new elements are again in need of analysis, 'they must be further examined by physiological research',[12] and so on. The hypothetical character of sensations also becomes clear from Mach's attempt[13] to put them in their proper place and to free science from their dominance. In

---

[9] 'Da aber in deisem *Namen* ('sensations') schon eine *einseitige Theorie* liegt, so ziehen wir vor, kurzweg von *Elementen zu sprechen . . .*' *Analyse der Empfindungen* (Jena: Gustav Fischer, 1900), p. 15, italics in the original.

[10] *Ibid.*, p.137.

[11] For details see Section 3 of my essay 'Against Method' in M. Radner and S. Winokur (eds.), *Minnesota Studies in the Philosophy of Science, volume IV* (Minneapolis: University of Minnesota Press, 1970).

[12] *Analyse der Empfindungen*, p. 20.

[13] *Notebooks*, vol. II, p. 16 from back.

debates with Hugo Dingler whom he praised in the Foreword to the seventh edition of his *Mechanik* he admits to being a 'non-empiricist' or a 'not-only-empiricist',[14] and he follows with interest Dingler's earlier attempts to eliminate experience from applied geometry. This partial anti-empiricism of Mach (which in truth is nothing but his universal criticism applied to the empiricist ideology) is a fascinating topic for research.

Now having *adopted* a *particular cosmological hypothesis* which seems plausible to him (elements are sensations) he *applies* it, and introduces the principle that science should contain only such concepts which can be connected with sensations. (It is useful to again compare this principle with the principle that scientific theories should be Lorentz-invariant). But it is interesting to see that he never relies on this 'empirical' criticism alone. Since the use of sensations is based on a hypothesis it is necessary to check it at every point by independent arguments. The objections to absolute space and to atomism are a case in point.

Absolute space is nonempirical, it cannot in any way be connected with sensations. Are there perhaps reasons which nevertheless force us to adopt it? There is Newton's bucket argument. This argument is invalid not because it makes appeal to things which are not related to sensations, but because it rests on a false assumption. Centrifugal forces, it is said, arise even if there is no material around with respect to which rotation can be asserted. This is untrue, for there are the fixed stars. If one now asserts that the fixed stars have no influence one asserts what has to be shown, viz., that centrifugal forces are not due to a particular relation to the rest of the universe. In this form the argument is circular *in addition to* yielding a metaphysical conclusion. Is it perhaps possible to decide the question by experiment? In order to advance in this direction Mach develops an alternative theory of his own in which inertial forces depend on the presence of matter and can be changed by the movement of large neighbouring masses (Friedländer's experiments were entirely in Mach's spirit). This alternative theory still survives and is now in the centre of discussion.

Mach's criticism of the atomic theory is another example of the way in which he combines his sense-data hypothesis with other, and more concrete, arguments. Again the metaphysical criticism is supported by difficulties of the existing atomic theories (reversibility objection; recurrence objection; stability of the atom). There is no reason to discredit Mach for his unwavering opposition to atoms.[15] For the mechanical atom of the nineteenth century has indeed disappeared from the scene.

---

[14] H. Dingler, *Die Grundgedanken der Machschen Philosophie*, p. 61, footnote.

[15] There has arisen the belief that Mach changed his mind in the last few years of his life and that he finally accepted atomism. This is based on a story told by Ernst Mayer, the head of the Radium Institute in Vienna at the time, who showed Mach a scintillation screen and reports him as saying, 'Jetzt glaube ich an die Atome'. There is no reason to doubt Mayer's story (Mayer was an

(**6**) To sum up: Mach develops the outlines of a knowledge without foundations. He introduces cosmological hypotheses as temporary measuring sticks of criticism. The cosmological hypothesis he appeals to most frequently assumes that all our knowledge is related to sensations. Using this hypothesis he criticizes physical theories such as the atomic theory and Newtonian mechanics. The cosmological criticism is strengthened by a more specific examination and by the suggestion of alternatives. In this way Mach achieves unity between science and philosophy that had been lost as the result of the dominance of Newton's physics and of his philosophy.

(**7**) The unity is achieved *and is at once lost again* by the empiricist successors of Mach. No longer is there a combined effort to criticize and to improve physics, nor is the critical attitude retained that was such an essential part of Mach's research. It is quite impossible to narrate and to explain all the developments which led from Mach through the Vienna Circle (plus the Berlin group, plus the Scandinavian groups, and so on) to the contemporary situation. Here is a fascinating field of study for the historian of ideas. All I can do in the present short note is to make a few scattered remarks and to formulate some questions.[16]

The first and most noticeable change is the transition from a critical *philosophy* to a *sense-data dogmatism:* elements are replaced by sensations not just temporarily, and as a matter of hypothesis, but once and for all. Sensations are regarded as the solid foundation of all knowledge. It would be interesting to know how this replacement came about and how it happened to be connected with the name of Mach. Dr Laudan has a simple suggestion and I am inclined to follow him (in the history of ideas the profound ideas are not always the best ones): Mach was either not read at all or read with little care. As a result the important distinction between basic philosophy and special hypotheses was never discovered, let alone understood. Besides, the end of the nineteenth century teems with sensationalistic philosophies. Of course, there were also more tolerant philosophies (Neurath; Carnap), but they never got beyond the abstract statement that the selection of an observation language (sense-data language; physical-thing language) was a matter of *choice*. There was never any analysis of the

honest man). But there is much reason to doubt the *interpretation* that is usually put on it (this has been pointed out to me by Dr Heinz Post of Chelsea). In itself the story is rather implausible. Why should Mach be convinced by a peripheral phenomenon such as the scintillation screen? There is also the fact that the optics which was published after the incident [the reference is to Mach's *Die Prinzipien der physikalischen Optik, historisch und erkenntnispsychologisch entwickelt* (Leipzig: Johann Ambrosius Barth, 1921), translated as *The Principles of Physical Optics: An Historical and Philosophical Treatment* (New York: Dover, and London: Methuen, 1926) (Ed.)] shows him as hostile to atoms as ever. Dr Post assumes that Mach's reaction was simply the reaction of a kind man to an enthusiastic (and rather deaf) demonstrator.

[16] Dr Laudan is now studying the history of the early Vienna Circle and he has already arrived at some very interesting results. The following brief sketch is influenced by some of these results, though it is written much more carelessly than Dr. Laudan would dare.

concrete steps and of the concrete arguments which might favour one choice rather than another.[17] *In practice* one assumed that a certain language was given and proceeded to discuss, reform, evaluate *theories* on its basis. And it did not last long, and the abstract principle of tolerance was abandoned also and was replaced by a more restrictive philosophy.[18]

Secondly, criticism of *science* is replaced by *logical reconstruction* which, in plain English, is nothing but a highly sophisticated brand of conformism. This development is partly connected with the first: if sensations are the foundation of knowledge, then the chains which connect, say, Maxwell's equations with the basis should be clearly exhibited and should be formulated as precisely as possible. Having found a reformulation of the chains one hoped to be able to rewrite the equations themselves, using the same precise language. Now there *did* exist a language which seemed to provide ready translations for mathematical formulas, viz., the language of *Principia Mathematica*. The logical reconstruction of science therefore amounted to the attempt to rewrite science in PM and to clearly exhibit all the links with the 'basis'.

Now an attempt of this kind may be understood in at least two different ways; *critically*: what cannot be reconstructed in this fashion must be eliminated; *conformistically*: what cannot be reconstructed in this way shows that the methods of reconstruction are faulty and must be revised. The critical version survived for some time – for had not Mach changed science by using a vague form of the verifiability principle and was not the theory of relativity the glorious result of this newly found philosophical method? But being isolated, and without help from more specific arguments (no one thought of inventing more concrete criticisms in addition to the arguments flowing from the reconstruction program), the verifiability criterion lacked strength to survive a showdown with science. In his famous debate with Planck Mach refused to revise his view of science just because it differed so much from the actual thing. Science has become a church, he said, and I have no intention <of being> a member of a church, scientific or otherwise. I therefore gladly renounce the title of a scientist: 'Die Gedankenfreiheit ist mir lieber'.[19] He argued from a strong position having both a plausible cosmological hypothesis and specific difficulties of the existing science on his side to prove his point. Not a single positivist was bold enough to wield the verifiability criterion in the same spirit. Driesch was criticized, yes, but this was a rather trivial case. There was no attempt to cut physics down to size in a similar manner. As a result the idea of a logical reconstruction

---

[17] I have discussed such arguments in a particular case in sections 5–10 of 'Against Method'.

[18] For a sketch of this development see my essay 'Explanation, Reduction, and Empiricism' in H. Feigl and G. Maxwell (eds.), *Minnesota Studies in the Philosophy of Science, volume III* (Minneapolis: University of Minnesota Press, 1962).

[19] *Zwei Aufsätze* (Leipzig, 1919).

became conformistic. The task was now to *correctly present* rather than to *change* science (it did not take long and this attitude was extended to the language of common sense also).

(**8**) This task in turn was soon transformed into problems of a different kind, some of which were no longer related to science at all. Trying to imitate the scientific procedure (or what one thought was the scientific procedure), one started with the discussion of *simple* cases. Now a simple case, in this context, was not a case that looked simple when viewed from inside science. It was a case that looked simple when formulated in the language of *Principia Mathematica*. In addition one concentrated on the relation to the evidence, omitting all those problems and aids which arise from the fact that every single statement of science is embedded in a rich theoretical net and chosen to fit this net in one way or another (today a 'simple' physical theory is a theory that is relativistically invariant). Now I do not wish to say that the properties of nets are not being discussed, for there is a large literature on precisely that point. All I wish to assert is that there exists an enterprise[20] which is taken seriously by everyone in the business where simplicity, confirmation, empirical content are discussed by considering statements of the form (x) $(Ax \rightarrow Bx)$ and their relation to statements of the form Aa, Ab, Aa & Ba, and so on and *this* enterprise, I assert, has nothing whatever to do with what goes on in the sciences. There is not a single discovery in this field (assuming there have been discoveries) that would enable us to attack important scientific problems in a new way or to understand better the manner in which progress was made in the past. Besides, the enterprise soon got entangled with itself (paradox of confirmation; counterfactuals; grue) so that the main issue is now its own survival and not the structure of science. That this struggle for survival is interesting to watch I am the last one to deny. What I *do* deny is that physics, or biology, or psychology can profit from participating in it.

(**9**) It is much more likely that they will be *retarded*.

This can be shown both theoretically, by an analysis of some rather general features of the present state of the program of reconstruction, and practically, by exhibiting the sorry shape of the subjects (sociology; political science) which have made a vulgarized version of the program their chief methodological guide. I shall restrict myself to a brief discussion of two theoretical difficulties.[21]

---

[20] Prominent textbook: I. Scheffler's *The Anatomy of Inquiry* (New York: Knopf, 1963).

[21] For more detailed analyses involving historical material the reader is referred to my essays 'Problems of Empiricism' in R. G. Colodny (ed.), *Beyond the Edge of Certainty* (Englewood Cliffs, NJ: Prentice-Hall 1965); 'Problems of Empiricism, II' in R. G. Colodny (ed.), *The Nature and Function of Scientific Theory* (Pittsburgh: University of Pittsburgh Press, 1970); 'Against Method'; and my criticism of Nagel in the *British Journal for the Philosophy of Science*, vol. 17, 1966, pp. 237–49.

The empirical model assumes a language, the 'observation language', that can be specified independently of all theories and that provides content and a testing basis for every theory. It is never explained how this language can be identified nor are there any indications how it can be improved. Carnap's rule that the language should be used by a language community as a means of communication and that it should also be observational in the vague sense of containing quickly decidable atomic sentences is clearly unsatisfactory. In the sixteenth and seventeenth centuries a language of this kind would have included devils, angels, incubi, succubi, absolute motions, essences, and so on. Now this fact, taken by itself, is no objection against the scheme *provided* the scheme permits us to overcome devils in a rational manner. This is not the case. An observation language is a final standard of appeal. There are no rules which would permit us to choose between different observation languages and there is no method that would show us how an observation language can be improved.[22] We may of course try to extract the devils from their observational surroundings and to formulate a demonic *theory* that can be interpreted and tested on the basis of a different observational idiom. This procedure is *arbitrary* (why should we not invert our procedure and test Dirac's theory of the electron on the basis of an Aristotelian observation language?). And it is *excluded* by the very same principles which make the model a judge of the empirical content of a notion (removing the observational embroideries of a concept changes the concept). The model is therefore incomplete at a very decisive point. And in order to remove the impression that it needs the *devil* to exhibit this incompleteness we should perhaps add that the transition from the physics of Aristotle to the physics of Galileo and Newton creates exactly the same problems. Here we see very clearly how the transition is achieved[23] and how complex are the arguments that bring it about (Galileo, for example, introduces a new observation language *in order to accommodate the Copernican theory*). Compared with this reality the double language model looks infantile indeed.

The second difficulty is that not a single theory ever agrees with all the known facts in its domain.[24] This being the case, what shall we make of the methodological demand that a theory must be judged by experience and must be rejected if it contradicts accepted basic statements? What attitude shall we adopt toward the various theories of confirmation and corrobora-

---

[22] In his essay 'Empiricism, Semantics and Ontology', reprinted in Leonard Linsky (ed.), *Semantics and the Philosophy of Language* (Urbana: University of Illinois Press, 1952), pp. 207–28, Carnap has discussed this feature with great insight. However, apart from distinguishing between external and internal problems he does not give any indication of how one should proceed.

[23] The arguments are presented in the references in 'Against Method'.

[24] For details see section 4 of 'Against Method'.

tion which all rest upon the assumption that theories can be made to agree completely with the known facts and which use the amount of agreement reached as a principle of evaluation? This demand, these theories, are now all quite useless. They are as useless as a medicine that heals a patient only if he is completely bacteria free. *In practice* they are never obeyed by anyone. Methodologists may point to the importance of falsifications – but they blithely use falsified theories; they may sermonize how important it is to consider *all* the relevant evidence, and never mention those big and drastic facts which show that the theories which they admire and accept, the theory of relativity, the quantum theory, are at least as badly off as are the older subjects they reject. In practice methodologists slavishly repeat the most recent pronouncements of the top dogs in physics, though in doing so they must violate some very basic rules of their trade.[25] Is it possible to proceed in a more reasonable manner?

(**10**) The answer is of course – yes. But the remedy needed is quite radical. What we must do is to replace the beautiful but useless formal castles in the air by a detailed study of primary sources in the history of science. *This* is the material to be analysed, and *this* is the material from which philosophical problems should arise. And such problems should not at once be blown up into formalistic tumours which grow incessantly by feeding on their own juices but they should be kept in close contact with the process of science even if this means lots of uncertainty and a low level of precision. There are already thinkers who proceed in this way. Examples are Kuhn, Ronchi, the late Norwood Russell Hanson, and especially Imre Lakatos who has almost turned the study of concrete cases into an artistic enterprise and whose philosophical suggestions can once more be used to transform the process of science itself. It is to be hoped that such a concrete study will return to the subject the excitement and usefulness it once possessed.

[25] In this, of course, they merely repeat Newton. The only difference is that the theories which Newton accepted were invented by himself.

7

On the limited validity of methodological rules

Translated by Eric M. Oberheim and Daniel Sirtes

> Ordnung ist heutzutage meistens
> dort, wo nichts ist. Es ist eine
> Mangelerscheinung.
>
> Order is nowadays mostly
> there where nothing is.
> It is a symptom of deficiency.
>
> Brecht

I INTRODUCTION

It is indubitable that the application of clear, well-defined, and above all 'rational' rules *occasionally* leads to results. A vast number of discoveries owe their existence to the systematic procedures of their discoverers.

But from that, it does not follow that there are rules which must be obeyed for *every* cognitive act and *every* scientific investigation. On the contrary, it is totally improbable that there is such a system of rules, such a logic of scientific discovery, which permeates all reasoning without obstructing it in any way. The world in which we live is very complex. Its laws do not lay open to us, rather they present themselves in diverse disguises (astronomy, atomic physics, theology, psychology, physiology, and the like). Countless prejudices find their way into every scientific action, making them possible in the first place. It is thus to be expected that every rule, even the most 'fundamental', will only be successful in a limited domain, and that the forced application of the rule outside of its domain must obstruct research and perhaps even bring it to stagnation. This will be illustrated by the following examples.

Footnotes added by the translators are marked with letters. All translators' additions to Feyerabend's own notes are contained within square brackets. Of a number of mistakes that we have found, we have noted those which seemed significant, and added some corrections where appropriate. We would like to thank M. Weber and N. Teubner for helpful suggestions, V. Hoesle for help with the translation of passages from Galileo, and the ZEWW, which partially funded this work.

2 THE FIRST EXAMPLE: NEWTON'S RULE IV

In science, it is a commonplace that the introduction of new ideas, hypotheses, and theories is allowed only after the orthodox point of view has run into contradiction with experience. New ideas must take the empirical difficulties of the accepted theories as a starting point. Without such difficulties, their invention is pointless and their discussion is an utter waste of time. Along these lines, E. Schücking and O. Heckmann write: 'According to our view, the continuous progress of the sciences was only possible because scientists thought it inadmissible to propose new theories *before* the empirical refutation of the older conceptions'.[1] T. S. Kuhn writes that 'invention of alternates is just what scientists seldom undertake ... The reason is clear. As in manufacturing so in science – retooling is an extravagance to be reserved for the occasion that demands it'.[2] This view can actually be traced back to Newton[3] who occasionally expresses himself with the help of the following rule: 'In experimental philosophy one is not to argue from hypotheses against propositions drawn by induction from phenomena. For if arguments from hypotheses are admitted against inductions, then the arguments of inductions on which all experimental philosophy is founded could always be overthrown by contrary hypotheses. If a certain proposition drawn by induction is not yet sufficiently precise, it must be corrected not by hypotheses but by the phenomena of nature, more fully and more accurately observed.'[4] In the *Principia*, the same rule appears as *Rule IV* and it now says: 'In experimental philosophy we are to look upon propositions inferred by general induction from phenomena as accurately or very nearly true, *notwithstanding any contrary hypotheses* that may be imagined, till such time as other phenomena occur, by which they may either be made more accurate, or liable to exceptions.'[A] This, then, is the *content* of the rule which we have chosen as our first example.

Its *function* is the following: it discredits ideas which contradict the orthodox point of view.

Newton already demanded that only those hypotheses about light which

---

[1] *World Models, Proceedings of the XIth Solvay Conference*, p. 1 [Our translation of Schücking and Heckmann].

[2] *The Structure of Scientific Revolutions* (Chicago: University of Chicago Press, 1962), p. 70. [English cited from 2nd edition, 1970, p. 76].

[3] As emphasized by Schücking and Heckmann.

[4] A. Koyré, *Newtonian Studies* (Chicago: University of Chicago Press, 1965), p. 269.

[A] Feyerabend does not provide a reference. The English translation has been taken from *Sir Isaac Newton's Mathematical Principles of Natural Philosophy and his System of the World*, translated by Andrew Motte, and revised by Florian Cajori (New York: Greenwood Press, 1962), p. 400. The italics are Feyerabend's.

are consistent with his own theory are permissible.[5] In the nineteenth century, one objection to the kinetic theory of matter was that it contradicts phenomenological thermodynamics (which was allegedly well-confirmed),[6] and this almost succeeded in pushing the kinetic theory out of science.[7] The arguments against hidden parameters in quantum theory have exactly the same structure,[8] except that the role of thermodynamics as an argumentative basis is now played by quantum theory. And even in cosmology, where speculation was allowed still greater freedom than in the sub-lunar parts of natural science, recently there are those who want to establish barriers to thinking in a similar way.[9] We see that Rule IV has a great influence on the concrete development of empirical research.[10] Therefore, it is important to determine its limits exactly, that is, the circumstances in which it obstructs research and perhaps even brings it to stagnation.

3 LIMITS OF THE RULE

For this purpose, consider a theory T which predicts a micro-process P'. In reality, micro-process P'' is the case, where $P'' \neq P'$. P'' triggers a macro-process M which can be observed by relatively simple means.

Thus one can say that M refutes theory T indirectly through P' and P''.

---

[5] All hypotheses about light 'should be conformable to my theories'. Letter to Oldenburg, 7 December 1675 (*The Correspondence of Isaac Newton I* (Cambridge: Cambridge University Press, 1959), p. 362). Also see sections 5 and 6 of the letter to Oldenburg, 11 June 1672, p. 177. For the role of Rule IV in the debate about light, see my article 'Classical Empiricism' in R. E. Butts and J. W. Davis (eds.), *The Methodological Heritage of Newton* (Oxford: Blackwell, 1970). Goethe saw the situation very clearly in his *Theory of Colours* [*Farbenlehre*].

[6] The arguments which show the contradiction are summarized in D. ter Haar, 'Foundations of Statistical Mechanics', *Reviews of Modern Physics*, vol. 27, 1955, pp. 289–338. For analysis, see Paul and Tatjana Ehrenfest's article in *Die Encyclopädie der Mathematischen Wissenschaften IV, 2*, Article IV, p. 32 (Leipzig, Teubner, 1911). See also Ernst Mach, *Wärmelehre* (Leipzig, 1897) and *Zwei Aufsätze* (Leipzig, 1912).

[7] This applies especially to the European continent. See von Smoluchowski, *Physikalische Zeitschrift*, vol. 13, 1912, p. 1070 and *Oeuvres 2* (1927) pp. 361ff. In England, the construction of models was an affair of national honour, see Stanley Goldberg's 'In Defense of the Aether: The British Response to Einstein's Special Theory of Relativity, 1905–1911', *Historical Studies in the Physical Sciences*, vol. 2, 1971.

[8] Discussion and literature can be found in sections 1 and 9 of my essay 'Problems of Microphysics' in R. G. Colodny (ed.), *Frontiers of Science and Philosophy* (Englewood Cliffs, NJ: Prentice-Hall, 1962).

[9] 'According to our view, it is good politics if one does not continue to pursue theories like those of Bondi, Gold, and Hoyle until one finds strong evidence for the continual generation of energy and momentum.' Schücking and Heckmann *World Models*, p. 2 [our translation].

[10] Philosophy of science will not be discussed here, because it has not developed any of its own ideas about science in the last decades. Instead, it directed its worshipping eyes toward science and has been prepared immediately to imitate even the slightest change in it.

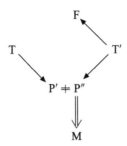

In addition, we assume that P′ and P″ are indistinguishable in principle by experimental means: there are laws of nature which forbid the possibility of distinguishing the two. Then T is refuted *de facto* by M without the scientist, who only uses T and 'the facts' in his considerations, ever being able to discover this circumstance. And M is pushed aside as a curiosity instead of contributing to the content of T.

Brownian motion of particles suspended in solution fulfils all the conditions just described. It is triggered by the molecular motion of the surrounding solution P″, but fluctuation phenomena in the solution and in the measuring instruments make it impossible experimentally to detect this triggering process, as well as the difference between P′ and P″.[11] *De facto*, the Brownian particle is a perpetual motion machine of the second kind, but this circumstance will always remain hidden to the experimenter who refuses to consult information drawn from other theories.

We now introduce a theory T′ which contradicts T, predicts P″, predicts the relationship between P″ and M, and also successfully explains other facts F which until then were unknown. Such a theory teaches us that M contradicts T. Thus, we refute T with the help of a fact whose relevance and refuting character can only be detected indirectly, i.e. through an alternative to T.

This situation also has its counterpart in the theory of Brownian motion. T′ is the kinetic theory of matter. The relationship between T′, P″, and M is contained in Einstein's theory of Brownian motion, where F, say, is the formula $x^2 \sim t$ which was investigated and confirmed by Perrin.

Now Rule IV forbids us to take alternatives to a theory seriously if the theory has not yet run into empirical difficulties. T′ is an alternative to T. *Seen by itself*, T is empirically flawless. The fact M which refutes T is either unknown, or is taken as a curiosity which obviously has nothing to do with

---

[11] Details and literature can be found in section IV of my article 'Problems of Empiricism' in R. G. Colodny (ed.), *Beyond the Edge of Certainty: Essays in Comtemporary Science and Philosophy* (Englewood Cliffs, N.J.: Prentice-Hall, 1965).

T.[12] Its relevance can only be demonstrated with the help of T', but this possibility is blocked. Rule IV orders us to wait for the first difficulty with T, but it is exactly this command that prevents the discovery of the first difficulty. This rule puts the cart before the horse and should be suspended, at least in our example.

What distinguishes our example from others? The only difference is in the *results* of the application of alternatives, that is, in a circumstance which only occurs *after* the violation has been carried out. *Before* this criminal act, T was a totally normal theory and M an isolated fact in no way different from other isolated facts. The theory T even has certain advantages: it is highly confirmed. It is interesting. It has applications in industry, cosmology, natural philosophy, and even journalism (heat death of the universe). *One must violate Rule IV in order to find out if the violation is appropriate.* This means, however, that a violation of Rule IV is both *permissible* and *imperative* in all circumstances.

Thus, for instance, it is permissible and imperative to introduce *hidden variables* into quantum theory and to study the consequences. After all, it is perfectly possible that certain difficulties of the theory, like for example the infinities in quantum field theory, may have their cause in the neglect of such parameters. The objection that hidden variables have not yet accomplished anything in quantum theory overlooks the fact that a complicated theory is not produced overnight. It took more than two thousand years before atomic theory became scientifically satisfactory. Furthermore, the orthodox theory has not gone very far, and it suffers from fundamental difficulties. In order to assess the difficulties properly, alternatives are urgently needed. And, after all, we are already much further than critics suppose. We are already at the stage of decisive experiments.[13] Thus nothing stands in the way of the further development of alternatives to the orthodox quantum theory.

The kinetic theory of matter and the theory of hidden variables were 'old' theories.[14] In other words, according to the view of influential groups, they were already overtaken by scientific progress. The example of Brownian motion (and other examples) show us that such an 'overtaking' is a provisional and temporary phenomenon which might be reversed by further research. The motive for such further research is often a certain

---

[12] That is the way Brownian motion was conceived of for a fairly long time. In addition, there was experimental evidence *against* conceiving of it as a molecular phenomenon: F. M. Exner, 'Notiz zu Browns Molekularbewegung', *Annalen der Physik*, vol. 2, 1900, p. 843. Exner determined that the motion is one order of magnitude less than the motion demanded by the law of equipartition.

[13] See J. F. Clauser, M. A. Horne, A. Shimony and R. A. Holt, 'Proposed Experiment to Test Local Hidden-Variable Theories', *Physical Review Letters*, vol. 23, 1969, pp. 880–4. The experiment is presently [1971] being prepared at Berkeley.

[14] This judgement applies to the kinetic theory on the European continent around the year 1860 and to hidden variables around the year 1930 (until around 1955).

metaphysical view which prefers a certain cosmology for personal reasons. Personal motives can be amplified on the grounds of our results by pointing out that alternatives increase the empirical content of orthodox views and facilitate their removal. This applies especially to those theories which have taken science by surprise and have now been left without rivals. Such theories, however, were once engaged in a battle with rivals and also emerged victoriously. But this only means that the mistakes, or apparent mistakes of the rivals, were discovered accidentally *before* their own mistakes. It certainly cannot be concluded that they are now free of mistakes themselves, or that their former opponents are exhausted or no longer fit for criticism. After all, they had only a limited time at their disposal and were never completely exploited. On the other hand, the theoretical criticism of the orthodox ideas was prematurely interrupted by pure accident. The revival of 'old' theories is therefore always reasonable and always holds the prospect of success.

Thus for example, it is reasonable and instructive to compare evolutionary theory, which dominates biology without qualification nowadays, with the theory of human development of *Genesis* and of *Enuma Elish*; and to re-confront the materialistic cosmology of modern physics and astronomy with the cosmology of Aristotle and the teachings of the Pimander. But here, one should take care to elaborate the latter cases in detail and to leave out Bultmann's attempts to castrate them.[15] Progress was often achieved by

---

[15] The attempt to demythologize [Entmythologisierung] 'sets out to clean away the offences which arise for the modern human being because he lives in a world view determined by science', writes R. Bultmann in 'Antwort an Karl Jaspers' in Jaspers and Bultmann's *Die Frage der Entmythologisierung* (Munich, 1954), p. 161. '*The actual problem is the hermeneutical one,*' Bultmann continues (p. 62), 'i.e. the interpretation of the Bible and of evangelism, in *that* way so that it can be understood as a word addressed to human beings'. In short, and less bombastically, the actual problem is a problem of *propaganda*, whereas ideas that disturb 'a modern human being' should be carefully left aside. Now the 'modern human being' 'is convinced that a corpse cannot step out of his grave and live again, that there are no demons, and no magic causal effects' (p. 62). Thus, 'the priest in his preaching and lessons' must hold the non-literal view of 'the bodily resurrection of Jesus, of demons, and of magic causal effects'. Such a view is not difficult for Bultmann because he believes 'that the myth is misunderstood if the reality of which he speaks is interpreted as 'empirical reality'' (p. 63, note) [our translation]. But such a misunderstanding is now out of the question. Myths often contain an empirical core which is confirmed by the 'facts' as well as, or sometimes even better than, the scientific ideas with which they competed in their heyday. (An impressive example is discussed by Trevor-Roper in his essay *The European Witch Craze* (New York: Harper Torchbooks, 1968). Also see the writings of Evans-Pritchard, Lévi-Strauss, de Santillana, and others.) And if such a core cannot be found, then it is still potentially there and can be brought to light through the development of the myth. But then, and now we come to a far more important point, who guarantees us that the ideas to which the 'modern human being' clings so much are correct, so that a 'priest in his preachings and lessons' only makes headway when he fully and wholly agrees with those ideas? Are those ideas exempt from criticism? And how has one been convinced of their correctness? Which concrete investigation has proven, or even made probable, that there are no demons? Although the assumption is part of science (even this is not totally sure; see the debates on

a 'criticism from the past' of the kind just described. After Aristotle and Ptolemy [Ptolemaios], the idea of the motion of the earth was considered antiquated once and for all ('Even the mere thought of such possibilities seems incredibly ridiculous', Ptolemy [Ptolemäus] writes[16]) but Copernicus, Kepler and Galileo revived it and led it to victory. The *Hermetic Writings* played no minor role in this process,[17] and they were still studied even by the great Newton.[18] Developments like these are not surprising when one considers that no idea is ever investigated in all of its ramifications, and that no point of view ever receives all the chances that it deserves. Theories are pushed aside and superseded by 'modern' ideas long before they have the opportunity to show their virtues in full light. And older opinions and myths only seem to be so completely without merits, either because one does not understand them,[19] or because their content is investigated by researchers whose knowledge of physics and astronomy is far beneath that of their originators.[20] When one takes the older opinions seriously,

the interpretation of quantum theory and also Jung's theory about the human soul), has anyone ever *directly* investigated it? Is it not rather the case that one assumes it because *other* parts of science are successful, and because one uncritically extends this success to the rest, without feeling the urge for independent verification? And how could one conduct such direct verifications? Is not the scientific method built in such a way that demons, if they exist, must always elude it? Except for if one begins with the assumption that there are demons, develops it in detail, and tries to find evidence with *its* help? But such a procedure demands that myths be taken literally, and rejects from the start, opportunistic attempts to castrate them.

[16] *Handbuch der Astronomie I* (ed. Manitius) (Leipzig, 1963), p. 18. [Our translation of Ptolemy].

[17] See for example the third dialogue of Bruno's *Aschermittwochsmahl*. For the role of the *Hermetic Writings* in the Renaissance, see Frances Yates, *Giordano Bruno and the Hermetic Tradition* (Chicago: University of Chicago Press, 1964), and the literature given there. For restrictions, see the articles by Mary Hesse and Edward Rosen in R. H. Steuwer (ed.), *Minnesota Studies in the Philosophy of Science, volume V* (Minneapolis: University of Minnesota Press, 1970).

[18] J. M. Keynes's 'Newton the Man' in his *Essays and Sketches in Biography* (London: Hart-Davis, 1933).

[19] In contrast with their predecessors (Galileo, Kepler, Newton, etc.), contemporary scientists are only very poorly acquainted with the history of their discipline, and they know older theories only in their rough outlines and often in a form distorted beyond recognition. This is especially the case in the physical sciences. In the *biological* sciences, one can sometimes find a certain historical education, and it is occasionally even utilized for the critique of 'modern' theories in exactly the same refined way in which Copernicus and Galileo confronted theories from antiquity with the school philosophies of their times.

[20] In view of the great precision of the classification of plants and animals which one finds in older myths and in non-Western tribes, and in view of the important role that astronomy plays in many myths (just think of the myths of the Polynesians), 'the wish arises that every anthropologist will also be an expert in mineralogy, botany, zoology, and even in astronomy,' writes Claude Lévi-Strauss, *La Pensée Sauvage* (Chicago, 1966), p. 45 [our translation of Feyerabend's German]. This condition is seldom fulfilled. Thus the astronomical, biological, and physical content of myths is lost through the usual translations and explanations, and is replaced by vague and senseless chatter out of which the so-called students of human nature infer the 'primitive' mental state of their inventors, while the primitivity is solely to be found in their own heads. See also G. de Santillana and H. von Dechend, *Hamlet's Mill* (Boston, 1969).

possibilities of interesting discoveries exist even in the centre of science. In this way, for example, some physiologists who belong to the materialistic school recently discovered that the phenomena of voodoo have a clear, though not too well-understood, *material* basis, and that studying it can contribute to the understanding of human functions, and maybe even to the revision of modern theories.[21] Let us not close ourselves to such possibilities on the basis of the conceited conception that the excellence of the present and the total absurdity of the past, or of non-Western forms of knowledge, are now already proven once and for all.

4 APPLICATION TO THE PROBLEM OF INDUCTION

The problem of induction consists in the question of how one can distinguish theories with the help of experience. This entails the tacit assumption that the theories thus distinguished offer more certainty than their less well-confirmed alternatives. Our considerations show us that such certainty can be severely disturbed by the appearance of insecure rivals. Instead of the rule 'use only highly confirmed and therefore certain theories in science', one should at least *occasionally* apply the rule 'test the certainty of the certain theories with the help of uncertain theories'. If one does this, one gives up the distinction between certain and uncertain theories, and with it all the theories of induction which try to explain and justify the difference between them. The problem of induction turns out to be a classic case of a pseudo-problem.[22]

After restricting Rule IV, it is just as impossible sensibly to speak of an 'approximation to the truth'. The kind of knowledge which emerges when one uses alternatives for criticism does not strive for such an ideal point of view. There is no theory that gradually comes forward and pushes aside its rivals. Every theory which has a tendency to such autocracy is immediately cut down to size by an alternative. Thus, we are dealing with a steady, growing sea of relatively inconsistent ideologies which force each other, as well as the consciousness of the scientist, to greater and greater articulation. Conclusive results are not achieved and no point of view is ever excluded from the debate forever. Plutarch and Diogenes Laërtius, but not Dirac and von Neumann, teach us how knowledge of this kind is to be *represented*. But the task of the scientist consists neither in 'the search for truth', nor in 'the glorification of God', nor in 'the systematization of observations'. The

---

[21] See C. R. Richter's 'The Phenomenon of Unexplained Sudden Death' in W. H. Gantt (ed.), *Physiological Bases of Psychiatry* (Springfield, 1958), pp. 112ff. Decisive preliminary studies were done by W. B. Cannon, *Bodily Changes in Pain, Hunger, Fear, and Rage* (New York, 1915), and "Voodoo' Death', *American Anthropologist*, NS vol. xliv, 1942. See also chapter 9 of C. Lévi-Strauss, *Structural Anthropology* (London: Jonathan Cape, 1967).

[22] See my article 'A Note on the Problem of Induction', *Journal of Philosophy*, vol. 61, 1964.

scientist's task is 'to make the weaker the stronger', as the Sophists expressed themselves, and thereby to keep our ideas in motion. This will become clearer in what follows.[23]

5 THE SECOND EXAMPLE: REFUTATION BY EXPERIENCE

The rule that a theory which contradicts experience must be excluded from science and replaced by a better theory was invented by Aristotle.[24] It was repeated emphatically by Newton,[25] and it plays an important role in the methodology of modern sciences.[26] Nevertheless, theories exist only because the rule is violated at every turn. *That is to say, there is not a single theory which is in agreement with all the facts in its domain.* And here I am not speaking about rumours, or about the results of sloppy procedures. The problem I am speaking about is created by experiments and measurements of the greatest precision and reliability.

At this point, it is practical to distinguish between numerical difficulties and qualitative failures. The first case is easy to describe. A theory makes a certain prediction, and the prediction is different from the numerical value experimentally obtained. Numerical difficulties occur every day in science.

As our example, we will take Newton's theory of gravitation. This unachieved idol of theoretical perfection was inconsistent with experience from the very beginning,[27] and 'numerous discrepancies between observation and theory' remain even today.[28] (We are speaking here of the non-

---

[23] Imre Lakatos, who writes well but can only read badly, represented the arguments above as if I had only described the *psychological* effect of alternatives, and he himself then emphasized their *logical* function in the process of refutation: 'Popper on Demarcation and Induction' in P. A. Schilpp (ed.), *The Philosophy of Sir Karl Popper* (La Salle, IL: Open Court, 1974), note 50 and text. The logical function was always my point and I have been trying to show it for more than ten years.

[24] *De Coelo* 306a7, 293a27, *De Generatione et Corruptione*, 325a13. *Analytica Priora* 43a14 shows the similarity with newer theories of falsification even more.

[25] See my article 'Classical Empiricism'.

[26] Even Putnam, in his relatively liberal article, 'Degree of Confirmation' in P. A. Schilpp (ed.), *The Philosophy of Rudolf Carnap* (La Salle, IL: Open Court, 1963), p. 772, demands that a theory should be maintained '*unless* it becomes inconsistent with the data'.

[27] For an overview, see chapters IV and V of Whewell's *History of the Inductive Sciences, volume II* (London, 1857. Reprinted by Frank Cass and Co. Ltd, 1967).

[28] Brower-Clemence, *Methods of Celestial Mechanics* (New York, 1961), p. v. See also R. H. Dicke, 'Remarks on the Observational Basis of General Relativity', in H-Y. Chiu and W. F. Hoffmann (eds.), *Gravitation and Relativity* (New York: Benjamin, 1964), pp. 1ff. A more exhaustive discussion of some of the difficulties of classical celestial mechanics can be found in chapters IV and V of J. Chazy, *La Théorie de la Relativité et la Méchanique Céleste, 1* (Paris, 1928). Just as clueless as all philosophers of science, Reichenbach praises Newton for his initial refusal to publish his theory of gravitation because it contradicted experience: 'Rather than set any theory, however beautiful, before the facts, Newton put the manuscript of his theory into his drawer' (*The Rise of Scientific Philosophy* (Berkeley and Los Angeles: University of California Press, 1959), pp. 101ff). And he continues: 'The story of Newton is one of the most striking illustrations of the method of modern science.' Reichenbach

relativistic domain.) Bohr's atomic model was not free from experimental defects for even a single moment, but Bohr and his followers continued to use it in their investigations.[29] The same applies to the special theory of relativity. This theory did not start from the facts, as is so often claimed,[30] but from certain theoretical difficulties.[31] One of its predictions was experimentally refuted after less than a year,[32] but Einstein did not take this circumstance very seriously.[33] A little later, the decisive refutation of the

overlooks the fact that the episode which he describes (and *falsely* describes, see L. T. Moore, *Isaac Newton. A Biography* (New York, 1962) chapters 9 and 11) is the exception and not the rule both in science and in Newton's life. Newton's methodology and his violations of it are described in my article 'Classical Empiricism'.

[29] For details, see chapters 2.2 and 3 of Max Jammer's *The Conceptual Development of Quantum Mechanics* (New York: McGraw-Hill, 1966). For analysis, see Imre Lakatos's 'Falsification and the Methodology of Scientific Research Programs' in I. Lakatos and A. E. Musgrave (eds.), *Criticism and the Growth of Knowledge* (Cambridge: Cambridge University Press, 1970), pp. 140ff., as well as John L. Heilbron and Thomas S. Kuhn's 'The Genesis of the Bohr Atom' in *Historical Studies in the Physical Sciences*, vol. 1, 1969. For the problems of the pre-Bohrian atomic model, see chapter 3 of John Heilbron's dissertation (Berkeley, 1964).

[30] R. A. Millikan, 'Albert Einstein on his Seventieth Birthday', *Reviews of Modern Physics*, vol. 21, 1949, pp. 343ff; R. B. Leighton's *Principles of Modern Physics* (New York, 1959), p. 5; *The Feynman Lectures on Physics, I* (Reading, MA: Addison-Wesley, 1963), pp. 15–23, and so on.

[31] See the first two pages of 'Zur Elektrodynamik bewegter Körper', *Annalen der Physik*, 1905. Einstein himself often claimed that the Michelson-Morley experiment first became known to him 'only after 1905' (R. S. Shankland, 'Conversations with Albert Einstein' *American Journal of Physics*, vol. 31, 1963, p. 48). For the whole problem of the empirical source of relativity theory, see G. Holton's 'Einstein, Michelson and the Crucial Experiment', *Isis*, vol. 60, 1969, pp. 133–97.

[32] W. Kaufmann, 'Über die Konstitution des Elektrons', *Annalen der Physik*, vol. 19, 1906, p. 487. Kaufmann interprets the measurement results as 'incompatible' with the Lorentz–Einstein theory. Lorentz expressed himself as follows: '. . . it seems very likely that we shall have to relinquish this idea altogether' (*Theory of Electrons*, 2nd edition, p. 213). Ehrenfest held the same opinion. 'Zur Stabilitätsfrage bei den Bucherer-Langevin Elektronen' *Physikalische Zeitschrift*, vol. 7, 1906, p. 302. Poincaré writes (*Science and Method* (New York: Dover Publications, 1952), p. 228): 'In their newer form [Kaufmann's experiments] *have shown Abraham's theory to be correct.* Accordingly it would seem that the Principle of Relativity has not the exact value we have been tempted to give it.' And he repeats (p. 286) 'A single experiment of Kaufmann's revolutionises at once Mechanics, Optics and Astronomy.' Poincaré never mentioned Einstein's treatise in his writings, and this empirical difficulty can be seen as at least one reason for this. See Stanley Goldberg's 'Poincaré's Silence and Einstein's Relativity', *British Journal for the History of Science*, vol. 5, 1970, pp. 73ff. After Kaufmann, Bucherer resumed the investigation of the question (*Physikalische Zeitschrift*, 1908, 1909) and arrived at the opposite result (see the footnote on p. 209 from Poincaré) after which Bestelmeyer again arrived at Kaufmann's results. And so on.

[33] *Jahrbuch der Radioaktivität und Elektronik*, vol. 4, 1907, p. 439. Einstein admitted the correctness of Kaufmann's calculations as well as the impossibility, at the time, of discovering systematic experimental mistakes. Despite this, he did not give up his own theory, since he thought that it was more satisfactory because it was more comprehensive than the existing alternatives. Furthermore, on the difficulties of the general theory of relativity, Einstein always emphasized 'The reason of the thing (*Vernunft der Sache*)' and did not think that 'verification through little effects' was very important. See the letter to Michele Besso and Karl Seelig, cited from G. Holton, 'Influences on Einstein's Early Work', *Organon*, vol. 3, 1966, p. 242 and Karl Seelig's *Albert Einstein* (Zurich, 1960), p. 271.

His indifference towards empirical results exposed itself especially in 1952 when theory by D. C. Miller found almost every scientist on the side of the theory.[34] Similar things apply to the general theory of relativity. Admittedly,

Freundlich's investigations about the deflection of light at the edge of the sun and about the red shift seemed to lead to values that did not agree with his theory. Born (letter to Einstein on May 4th, 1952 cited in *The Born-Einstein Letters* (New York: Macmillan, 1971), p. 190) sees the situation as follows: 'It really looks as if your formula is not quite correct. It looks even worse in the case of the red shift; this is much smaller than the theoretical values towards the centre of the sun's disk, and much larger at the edge. What could be the matter here? Could this be a hint of non-linearity?' Einstein (letter of 12 May 1952, p. 192) replies, 'Freundlich ... does not move me in the slightest. Even if the deflection of light, the perihelial movement or line shift were unknown, the gravitation equations would still be convincing because they avoid the inertial system (the phantom which affects everything but is not itself affected). It is really strange that human beings are normally deaf to the strongest arguments while they are always inclined to overestimate measuring accuracies.'

One has to keep all of this in mind in order to see Feigl's oft-repeated report of Einstein's 1920 Prague lecture in the right light. Feigl writes (*Minnesota Studies in the Philosophy of Science, volume V* (Minneapolis: University of Minnesota Press, 1970), p. 9), 'If ... Einstein relied on "beauty", "harmony", "symmetry" and "elegance" in constructing ... his general theory of relativity, it must nevertheless be remembered that he also said (in a lecture in Prague, 1920 – I was present then as a very young student): "if the observations of the red shifts in the spectra of massive stars don't come out quantitatively in accordance with the principles of general relativity, then my theory will be dust and ashes"'. I do not doubt Herbert Feigl's report, but I doubt that Einstein took this assertion very seriously. His estimation of Freundlich's results, just cited, shows this very clearly. And the objection that here one is concerned with the metaphysical degeneration of an aged empiricist can immediately be dismissed with reference to the 1907 article (cited at the beginning of this footnote) in which one finds exactly the same attitude. See also G. Holton's article 'Mach, Einstein, and the Search for Reality', *Daedalus*, vol. 97, 1968, pp. 636ff., especially p. 651.

Other historical remarks which Feigl draws on to support his empiricism can be dismissed in a similar way. Feigl claims that it is possible to determine and write down empirical facts without recourse to theories: 'Among the countless examples that are ready at hand, let me mention just a few: the phenomenon of Brownian motion can be described independently of the explanations given by Einstein and Smoluchowski ...' (p. 8). This is indeed the case, but the most precise descriptions which were available before Einstein *were wrong*, and Einstein's theory was required in order to find better methods of measurement. See footnote 12 and text of the present article, as well as footnotes 90–2 and text.

[34] I call this refutation 'decisive' because it was executed as well as that of Michelson and Morley at that time, despite *abstract* doubts. Lorentz studied Miller's investigations for many years, but could not find the mistake which he *hoped* to find. The explanation was discovered only more than twenty-five years after Miller's experiment. See R. S. Shankland, 'A New Analysis of the Interferometer Observations of Dayton C. Miller', *Reviews of Modern Physics*, vol. 27, 1955, pp. 167ff., as well as, 'Conversations with Albert Einstein', *American Journal of Physics*, vol. 31, 1963, pp. 47–57, especially p. 51 and footnotes 19 and 34. See also the discussion of the Conference on the Michelson–Morley experiment, reported in *Astrophysical Journal*, vol. 68, 1928, pp. 341ff.

In 1922, Born had already remarked on Michelson's experiment (letter to Einstein of 6 August 1922, pp. 73ff.): 'The Michelson experiment is one of those which seem definitely *a priori*', and he added to this remark: 'When I was in the United States in 1925/6 Miller's measurements were still frequently being discussed. I therefore went to Pasadena to see a demonstration of the apparatus on top of Mt Wilson. Miller was a modest little man who very readily allowed me to operate the enormous interferometer. I found it very shaky and unreliable; a tiny movement of one's hand or a light cough made the interference fringes so unstable that no readings were possible. From then on I completely lost faith in Miller's results. I knew from my visit to Chicago in 1912 that Michelson's own apparatus was very reliable and his measurements accurate.'

the theory was very successful in a certain respect.[35] But here also, there was a series of failures, beside the striking and exciting successes. The theory could in no way explain the nodal lines of Mars (5″) and Venus (10″),[36] and now there are renewed difficulties, mainly deriving from new calculations of Mercury's orbit by Dicke and others.[37]

These examples can easily be proliferated. As a matter of fact, we can confidently claim that for every theory which is not totally empty and uninteresting, there are domains in which it quantitatively contradicts experience, and other domains in which it fails qualitatively. It is not easy to discover such domains because there are many methods to make them invisible. Textbooks do not normally mention the difficulties. Quite to the contrary, they perceive the represented theory with the same pious faith with which a conservative catholic notes the Immaculate Conception of the Virgin Mary. *Original treatises* are no less misleading because they disguise the *way* in which the difficulties expose themselves, and thereby garble their character. Thus, there is a need for quite extensive investigations in order to dig up the omnipresent fact of the failures of even our best theories, and to make them believable.

What has just been said about the numerical difficulties applies to the *qualitative failures* of a theory to an even greater extent. There are a great many of them, but they are unknown. The following examples (which can also easily be proliferated) are notable.

According to Newton, light consists of *rays* of different refrangibility, which have only very little lateral extension. Light rays can be separated, unified, refracted, and partially weakened by absorption, but their inner constitution can never be changed. If we consider that the surface of a mirror possesses unevennesses which are greater than the lateral extension of the rays, then we see immediately that the ray theory is unable to explain mirror images. If light consists of rays, then the mirror should behave like a rough surface and should look like a wall.[38] Newton admits

---

[35] Certain measurements of the deflection near the sun's disk and of Mercury's orbit are examples. However, one has to consider that the calculation of Mercury's orbit makes essential use of Newtonian Mechanics. Over 5000″ of the motion of Mercury's perihelial movement was calculated purely classically, and the formalism of the general theory of relativity was thrown into gear only for the remaining 43″.

[36] Chazy *ibid.*, vol. I, p. 230.

[37] Dicke, 'Remarks' in Chiu and Hoffmann, *Gravitation*.

[38] Conversely, Galileo and, even earlier than he, Oresme, inferred that the moon cannot be a mirror. See Galileo, *Dialogue Concerning The Two Chief World Systems* (trans. Stillman Drake) (Berkeley and Los Angeles: University of California Press, 1953), pp. 71ff., as well as Nicole Oresme, *Le Livre du Ciel et du Monde* (eds. A. D. Menut and A. J. Denomy) (Madison, 1968), p. 457: 'Nor do we see the sun's light on the moon as in a mirror, for we should then not see the moon as we do; rather, the sun would appear in only a small portion of that part of the moon which seems lighted to us, and at times it would appear in no part at all; and it would be seen in different parts at different times ...'

this, but retains his theory and removes the difficulty with the help of an *ad hoc* hypothesis: 'the Reflexion of a Ray is effected, not by a single point of the reflecting Body, but by some power of the Body which is evenly diffused all over its Surface'.[39]

Newton removes the qualitative discrepancy between theory and fact with the help of an *ad hoc* hypothesis. In other cases, one regards this manoeuvre as unnecessary. One retains the theory *and tries to forget its disadvantages*. An example is the attitude toward Kepler's rule according to which an object observed through a lens is seen at the intersection of the rays hitting the eye.[40] The rule predicts that one will see an object lying at the focal point at an infinite distance. 'But on the contrary,' writes Isaac Barrow, Newton's teacher and predecessor in Cambridge,[41] 'we are assured by experience that [a point situated close to the focus] appears variously distant, according to the different situations of the eye ... And it does almost never seem farther off than it would be if it were beheld with the naked eye; but, on the contrary, it does sometimes appear much nearer... All which does seem repugnant to our principles. But for me', Barrow continues, 'neither this nor any other difficulty shall have so great an influence on me, as to make me renounce that which I know to be manifestly agreeable to reason'.

Barrow *mentions* the difficulty and *emphasizes* that he wants to maintain the theory anyway because it is 'manifestly agreeable to reason'. The super-empiricists of the twentieth century act in exactly the same way, but are much more inclined to hide the difficulties which they meet.

For example, the classical electrodynamics of Maxwell-Lorentz implies that the motion of a free particle is self-accelerated.[42] If one plugs the self-energy of the electron into the calculation, then one gets diverging expressions for point charges, while finitely extended charges amount to the relativistically demanded transformation properties for particles only if

---

[39] *Optics* (New York, 1952), p. 266. For discussion of this example, also see my 'Classical Empiricism'.

[40] See Kepler's *Ad Vitellionem Paralipomena* in J. Kepler, *Gesammelte Werke II*, edited by order of the Deutsche Forschungsgemeinschaft and the Bayrische Akademie der Wissenschaften (Munich, 1939), p. 72. A detailed discussion of the history of the rule and its influences can be found in Vasco Ronchi's *Optics: The Science of Vision* (New York: New York University Press, 1957). The terrible influence of the rule on the craft of making glasses is described by von Rohr, *Das Brillenglas als optisches Instrument* (Berlin, 1934), pp. 1ff. See also Gullstrand's Supplements to Part I of the English translation of Helmholtz's *Physiologische Optik* (New York, 1962), pp. 261ff.

[41] *Lectiones XVIII Cantabrigiae in Scholis publicis habitae in quibus Opticorum Phenomenon genuinae Rationes investigantur ac exponentur* (London, 1669), pp. 125ff. Berkeley uses this passage in his attack on the traditional 'objectivistic' optics: *Towards a New Theory of Vision* in *Works* (ed. Frazer) (London, 1901), pp. 137ff.

[42] See D. K. Sen, *Fields and/or Particles* (Toronto: Ryerson Press, 1968), p. 10. In the non-relativistic approximation, the factor is $\exp(3/2\ mc^3/e^2)$ in which m is the observed mass of the particle.

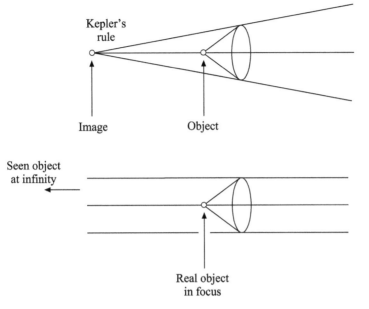

one introduces highly artificial and untestable pressures and shearings inside the electron.[43] The difficulty remains in quantum theory, but there it is partially hidden by the procedure of renormalization. This procedure consists of crossing-off the results of certain calculations and replacing them by descriptions of the actually observed conditions. With this, one implicitly admits that the theory involves insurmountable qualitative difficulties, but one formulates this confession so as to give the impression that a new theory has been discovered, and not that the old theory is refuted.[44] No wonder that philosophically uneducated authors get the impression, 'that all the evidence proves with merciless definiteness the conformity of the unknown interactions with the fundamental quantum law'.[45]

The history of the so-called 'Copernican Revolution' is an extremely

---

[43] See W. H. Heitler's *The Quantum Theory of Radiation*, 3rd edition (Oxford: Oxford University Press, 1954), p. 31.

[44] Besides this methodological objection, the theory of renormalization is also exposed to factual difficulties. For the whole complex of problems, see *The Discussions of 12th Solvay Conference, The Quantum Theory of Fields* (New York and London, 1962), especially the contributions from Heitler and Feynman. Born (*The Born-Einstein Letters*, p. 105) bluntly calls renormalization an 'almost grotesque trick'.

[45] L. Rosenfeld in S. Körner (ed.), *Observation and Interpretation* (London: Butterworth, 1957), p. 44. *[Feyerabend has slightly paraphrased here. The citation actually reads: 'Unfortunately for them, all the evidence points with merciless definiteness in the opposite direction: however strange they may be in other respects, all the processes involving the unknown interactions invariably conform to the fundamental quantum law.']

surprising example of a qualitative difficulty and how it was gradually overcome.

This fascinating and highly instructive event has been variously interpreted in the past. For example:

(1) As the transition from a metaphysical period, which was dominated by speculation, to a period of observation. Briefly but drastically put, the predecessors of Copernicus read their astronomy in *books*. Copernicus, on the other hand, directed his gaze at the *sky* and discovered the true astronomy. This interpretation has almost no followers today.[46]

(1a) As the transition from a complicated and over-loaded theory to a much simpler theory. This interpretation, which is also popular nowadays, meets with difficulties.[47]

(2) As the transition from a refuted theory to a new point of view which is able to cope with the difficulties of the refuted theory. This is the interpretation of Schücking and Heckmann.

(3) As an historical illusion caused by insufficient knowledge of the physics and astronomy of the Middle Ages. This is (roughly speaking) Duhem's position.[48]

The judgement made by the immediate successors of Copernicus and by his contemporaries differs depending on whether the theory is perceived as a calculation scheme, or as a new physical description of the universe. In the first case, the motion of the earth is purely fictitious, like a co-ordinate transformation which one executes in order to simplify the solution to mathematical problems, without thereby assuming the existence of special physical processes.[49] This kind of theory slowly penetrated into school astronomy[50] by way of Reinhold's *Prutenic tables*.[51]

---

[46] That it could arise at all can probably be traced back to a lack of knowledge about the arch-empiricist Aristotle (empiricists often begin to dream when they consider the *history* of their discipline), as well as the sad impression which the Aristotelians of the sixteenth and seventeenth century made on their contemporaries. For that, see especially L. Olschki's *Geschichte der Neusprachlichen Wissenschaften Literatur*, 3 volumes (reprinted: Vaduz, 1965).

[47] This interpretation only refers to so-called 'mathematical astronomy', i.e., to the calculation of the right ascension and declination without considering the change of luminous power and of dynamical laws. It is investigated in R. Palter, 'An Approach to the History of Early Astronomy', *Studies in History and Philosophy of Science*, vol. 1, 1970.

[48] For this interpretation, see section 9 of Agassi's *Towards an Historiography of Science*, Beiheft 2 to *History and Theory*, 1963.

[49] This interpretation of the motion of the stars was generally accepted in late antiquity and the Middle Ages. See P. Duhem, *To Save the Phenomena* (Chicago: University of Chicago Press, 1969).

[50] One can see how slowly if one investigates the textbooks of the time. This can be found in Francis R. Johnson, 'Astronomical Textbooks in the 16th-Century' in *Science, Medicine and History: Essays in Honour of Charles Singer, volume 1* (Oxford: Oxford University Press, 1953), pp. 285ff., and in Lynn Thorndike, *A History of Magic and Experimental Science, volumes V and VI* (New York: Columbia University Press, 1941).

[51] *Prutenicae Tabulae Coelestium motuum* (Wittenberg, 1551).

In the second case, one ascribes two motions to the earth:[52] it rotates around its axis, and revolves around the sun. This form of the theory was exposed to objections in the sixteenth and early seventeenth centuries which were serious enough to render it considered refuted.[53] We distinguish dynamical objections which were mainly directed against the rotation of the earth from optical objections. The dynamical objections tried to show that the idea of the rotation of the earth is in contradiction with certain simple and widely known *facts*, presuming that one combines the facts with a highly confirmed and plausible *theory* of motion. They have the same structure as the arguments which today want to prove the motion of the earth with the help of Foucault's pendulum, except that Aristotelian dynamics takes the place of Newtonian dynamics. The optical objections are directed at the contradiction between the factual change in brightness of the planets, mainly of Mars and Venus, and the change in brightness which is to be expected given the change of their distance from the earth in the Copernican system.

Galileo formulates the objections in the following way:

> If we move together with the earth with such a great velocity towards the east, then all other objects which are detached and separate from the earth would have to appear to move with exactly the same velocity towards the west. And therefore the birds of the air and the clouds, which cannot follow the motion of the earth, would have to remain in the west. Moreover, an object dropped from a high place, like a stone dropped from the top of a tower, would not arrive at the foot of the tower. This is because during the time in which the stone is on its perpendicular path through the air towards the earth, the earth would rush away beneath it and move to the east, so that as a result the stone would fall to earth in a place far away from the foot of the tower, just as a stone dropped from the mast of a fast-moving ship does not arrive at the foot of the mast, but more towards the stern. One would also recognize this even more clearly in the case of objects thrown perpendicularly upwards, which would, on their return, hit the earth a great distance from the position from which they were thrown. And also the arrow which one shoots skywards would not fall down near the bowman.[54]

---

[52] *Three* according to Copernicus, see *Commentariolus* (translated by Rosen) in *Three Copernican Treatises* (New York: Dover, 1959), pp. 63ff. The third motion can be left out if one assumes that the axis of the earth remains parallel to itself throughout the earth's revolution around the sun. Kepler, *Weltgeheimnis* (ed. Kaspar) (Munich, 1936), chapter 1, note 17.

[53] The theory, said Galileo in the *Saggiatore*, is 'surely false'. *The Controversy on the Comets of 1618* (eds. Drake and O'Malley) (Philadelphia, 1960), p. 185.

[54] *Trattato Della Sfera* in *Opera II* (ed. Favaro) (Edizione Nazionale), p. 224. This treatise is still written in a totally Ptolemaic spirit, and even keeps to the structure of Ptolemaic textbooks. It is interesting to note that stones dropped from the mast of a moving ship do not land at its base. Later, Galileo turned 'the facts' around and used the condition that stones land at the base of the mast even in quickly moving ships as an argument *against* Aristotle, which shows that we are dealing with a *thought experiment* which can be adapted to the necessities of the result to be reached. *Real* experiments were executed, but their results were anything but conclusive. See A. Armitage, 'The Deviation of Falling Bodies', *Annals of Science*, vol. 5,

The experiences which contradict the annual movement [of the Earth] have ... even greater force [than the dynamic arguments against the rotation which were just mentioned].[B] Mars, when it is close to us, would have to look sixty times as large as when it is most distant. No such difference is to be seen. Rather, the planet, in opposition to the sun, is only four or five times as large as at the time of conjunction, when it becomes hidden behind the rays of the sun.

Another and greater difficulty is made for us by Venus, which, if it circulates around the sun as Copernicus says, would be now beyond it and now on this side of it, receding from and approaching toward us by as much as the diameter of the circle it describes. Then when it is beneath the sun and very close to us, its disc ought to appear to us a little less then forty times as large as when it is beyond the sun and near conjunction. Yet the difference is almost imperceptible.[C] [55]

1941–7, pp. 342ff., and A. Koyré, *Metaphysics and Measurement* (Cambridge: Cambridge University Press, 1968), pp. 89ff. From *this* side, no help was to be expected in the sixteenth and early seventeenth century.

This argument has a long and interesting history itself. It is to be found in Aristotle, *De Coelo*, 296b22. Ptolemy uses it in chapter 7 of the first book of his main work. Copernicus also uses it in the seventh chapter of *De Revolutionibus*, but he tries to dismiss it in the eighth (see *Kreisbewegungen* (ed. Menzzer) (Thorn, 1879), pp. 18ff). The role of the argument in the Middle Ages is described in chapter 10 of M. Clagett's *The Science of Mechanics in the Middle Ages* (Madison, 1959).

[B] The actual English text reads: 'The experiences which overtly contradict the annual movement are indeed so much greater in their apparent force' and this part of the citation can be found in *Dialogue Concerning The Two Chief World Systems*, p. 328.

[C] Feyerabend has taken great liberties in his German translation. The actual English text on p. 334 reads: 'For if it were true that the distances of Mars from the earth varied as much from minimum to maximum as twice the distance from the earth to the sun, then when it is closest to us its disc would have to look sixty times as large as when it is most distant. Yet no such difference is to be seen. Rather, when it is in opposition to the sun and close to us, it shows itself as only four or five times as large as when, at conjunction, it becomes hidden behind the rays of the sun'.

[55] *Dialogue Concerning The Two Chief World Systems*, p. 334. Galileo uses the relationships 1:8 (Mars) and 1:6 (Venus) – pp. 321ff. – for the variation of the distance from the earth and allows the luminous power to vary in relationship to the observable surfaces. The numbers for the distances are obtainable from *Commentariolus* or from the *Revolutions* if one takes into account the first epicycle of the first and the eccentricity of the second. In the *Commentariolus*, the values for Mars are as follows: radius of the earth's orbit, 25; radius of the deferent, 38; radius of the first epicycle, 5 (for these numbers see Rosen, (ed.), *Three Copernican Treaties*, p. 74 and p. 77). Therefore, the relationship is 50 + (38−25) + 5 / (38−25) −5 ∼ 8. In the *Kreisbewegungen* (*Revolutions*), the values are (Menzzer, p. 330): radius of the earth's orbit, 6580; radius of Mars' orbit, 10000; eccentricity of Mars' orbit, 1460. Claudius Ptolemy, *Handbuch der Astronomie II* (ed. Manitius) (Leipzig, 1963), p. 198 and p. 197 gives for the relationship of excenter-radius : epicycle radius : eccentricity 60 : 39p30′ : 6, and thus a similar value for the relationship between perigee distance to apogee distance.

In addition to Galileo, the problem is mentioned by the oft-defamed Osiander in his notorious preface to Copernicus' main work: 'For these hypotheses need not to be true nor even probable. On the contrary, if they provide a calculus consistent with the observations, that alone is enough. Perhaps there is someone who is so ignorant of geometry and optics that he regards the epicycle of Venus as so probable, or thinks that it is the reason why Venus sometimes precedes and sometimes follows the sun by forty degrees and even more. *Is there anyone who is not aware that from this assumption it necessarily follows that the diameter of the planet at perigee should appear more than four times, and the body of the planet more than sixteen times, as*

And he describes the situation of the Copernican theory in the following way:

You wonder that there are so few followers of the Pythagorean opinion, whereas I am astonished that there have been any up to this day who have embraced and followed it. Nor can I ever sufficiently admire the outstanding acumen of those who have taken hold of this opinion and accepted it as true; they have through sheer force of intellect done such violence to their own senses as to prefer what reason told them over that which sensible experience plainly showed them to the contrary. For the arguments against the whirling of the earth [see first citation above] which we have already examined are very plausible, as we have seen; and the fact that Ptolemaics and Aristotelians and all their disciples took them to be conclusive is indeed a strong argument of their effectiveness. But the experiences which overtly contradict the annual movement [second citation] are indeed so much greater in their apparent force that, I repeat, there is no limit to my astonishment when I reflect that

great as at the apogee? Yet this variation is refuted by the experience of every age'. *Kreisbewegungen* (ed. Menzzer) p.1 (my emphasis). [The English translation of Osiander used here is that by Edward Rosen from J. Dobrzycki (ed.), *Nicholas Copernicus. Complete Works, volume II* (Poland: Polish Scientific Publishers and London: Macmillan, 1978), p. xvi]. The part emphasized, which is suppressed by all of Osiander's critics, puts his instrumentalism and the instrumentalism of his contemporaries in a new light. We know that this philosophy was held for theological and tactical reasons (Osiander's letter to Rheticus from 20 April 1541, reprinted in K. H. Burmeister, *Georg Joachim Rheticus III* (Wiesbaden, 1968), p. 25), and also because it fitted well in the astronomical tradition (letter of the same day to Copernicus, translated in Duhem, *To Save to Phenomena*, p. 68). But Osiander's instrumentalism also had physical reasons on its side, namely, exactly the same arguments that Galileo published in his *Dialogue* almost a century later. After all, it had been shown that the Copernican theory, interpreted realistically, is refuted by the facts. One has to bear this in mind if one takes to heart Karl Popper's bombastic, sentimental article, 'Three Views Concerning Human Knowledge' (in *Conjectures and Refutations* (New York: Basic Books, 1962), pp. 97ff). (The same goes for Popper's *evaluation* of quantum theory; here as well as there, he spies on instrumentalistic ideas with a Hooverian devotion, without mentioning, in even a single word, the physical arguments for these ideas.) Or, if one reads Dean White's *A History of the Warfare of Science with Theology, volume I* (New York: Dover Books), p. 123: 'But Osiander's courage failed him: he dared not launch the new thought boldly. He wrote a grovelling preface, endeavouring to excuse Copernicus for his novel idea ...' Would it have been better if Osiander were to have courageously introduced a theory for whose falsity he had arguments which even Galileo thought were very convincing?

Osiander's argument was discussed by Bruno (*Das Aschermittwochsmahl, Werke I*, translated by Ludwig Kuhlenbeck (Leipzig, 1904), pp. 87ff.) and thoroughly rejected: 'From the apparent magnitude of a luminous body, one can never infer its real magnitude or distance' [our translation]. Galileo had a greater interest in this argument since he tried to show that the telescope overcomes the previously described difficulty. The *source* of this argument can most likely be found in the fact that the *Ptolemaic* theory often did not give the right diameter, but still was maintained as an instrument for the calculation of longitudes and latitudes. Thus, the theory was not a description of actually occurring events ('one does not find such paths in the sky', Bicard writes in his *Quaestiones Novae in libellum de Sphaera Joannis de Sacro Bosco* (Paris, 1552), cited in Duhem, *To Save the Phenomena* p. 74., 'we only introduce it to show to those who learn astronomy how one can save the motions of the heavenly bodies' [our translation]). This point of view was easily transferable to the problems with the Copernican theory.

Aristarchus and Copernicus were able to make reason so conquer sense that, in defiance of the latter, the former became mistress of their belief.[56]

Thus we have here a fourth interpretation of the transition from the Middle Age cosmology to the cosmology of Copernicus. Copernicus neither offered new facts which could provide an inductive basis for his own ideas, nor did he know of observations which would refute Ptolemy but which agreed with his own theory.[57] On the contrary, both theories, Copernicus' as well as Ptolemy's, have difficulties, the first perhaps even greater than the second. Despite this, the Copernicans did not give up, but insisted on the correctness of their theory. Led by the 'vividness of their spirit', they did violence to the senses and led reason to victory.

6 METHODOLOGICAL REMARKS

These examples, which can easily be proliferated, show us that scientific practice is only seldom in accordance with logical and epistemological demands. As a matter of fact, there is not a single principle which was not repeatedly violated in the history of the sciences, including even such 'basic' and 'evident' principles as the principle of consistency. Scientific theories, as they occur in the history of science, are not only uncertain and always exposed to refutation, but are already refuted in every moment of their existence: they are in numerical difficulties. They contain deep qualitative mistakes. *Ad hoc* hypotheses patch up gaps in the proofs and cracks in the connection with facts. And internal contradictions are almost never avoided. We do not have proud cathedrals standing before us, instead we have dilapidated ruins, architectural monstrosities whose precarious existence is laboriously prolonged through ugly patch-work by their constructors. *This* is the scientific reality. *It* is not spoken of in the theory of knowledge or in the philosophy of science. How is this discrepancy to be explained? And what consequences for the evaluation of methodological rules follow from it? This is the problem to which I would like to sketch an answer in this article.

The first reason for this discrepancy lies in the deceptive way in which scientific theories (since Euclid) are represented and discussed within science itself. Von Neumann's book, *Mathematische Grundlagen der Quantenmechanik*,[58] which has many successors today, is an excellent example. Step by step, concepts are introduced and clarified. Problems are posed and solved. And thus, a building of awe-inspiring clarity and precision, resting on

---

[56] Cited from Galileo Galilei *Sidereus Nuncius. Nachricht von neuen Sternen* (ed. H. Blumenberg) (Sammlung Insel, 1965), p. 208. [This citation is incorrect. The text is not to be found in *Sidereus Nuncius*, but in *Dialogue Concerning The Two Chief World Systems*, p. 327–8.]

[57] '... Ptolemy's theory of the planets and that of most other astronomers [are] in agreement with the numerical data ...' writes Copernicus in *Commentariolus* (ed. Rosen) p. 57.

[58] (Berlin: Springer, 1932).

simple and easily understandable foundations, gradually arises. It seems that the confusion of the older quantum theory is overcome once and for all. But let us ask ourselves what relationship this sublime figment of the imagination has to reality! That is, let us ask ourselves (1) how the *assignment* of the hypermaximal operators of the theory are made to the concrete measurement apparatuses (2) how one makes *predictions* after one has accomplished the assignment, and (3) how one *tests* these predictions. And we will see that the impression of perfection can only arise because we thought that some peaceful corner of the ruin of 'quantum mechanics' was the whole building. The answer to (1) is that 'for some observables, in fact for the majority of them (such as xyp), nobody seriously believes that a measuring apparatus exists'.[59] In practice, one still draws on the older correspondence principle here. What can be said about (2) is that certain important predictions within the scope of the theory are only possible when one changes *other* laws arbitrarily.[60] Third, testing predictions is largely restricted by the way in which the coincidence counter classifies events. And let us not forget the qualitative problems of the theory which were mentioned in section 5. The clarity, precision, and simplicity which we find in von Neumann's book, and in other textbooks of theoretical physics, is thus, like all earthly perfections, *mere illusion*.[61] It is this illusion to which

---

[59] E. P. Wigner, 'The Problem of Measurement', *American Journal of Physics*, vol. 31, 1963, p. 14. 'Quantum mechanics', writes Schrödinger about this situation, 'claims that it deals ultimately and directly with nothing but actual observations, since they are the only real thing, the only source of information, which is only about *them*. The measurement is carefully phrased so as to make it epistemologically unassailable [...] But what is all this epistemological fuss for, if we have not to do with actual, real findings 'in the flesh', only with imagined findings?' (*Nuovo Cimento*, 1955, pp. 7ff.) Bridgman writes that quantum theory seems, *prima-facie*, to be a thoroughgoing operational theory: 'This end is achieved by labelling some of the mathematical symbols "operators", "observables", etc. But in spite of the existence of a mathematical symbolism of this sort, the exact corresponding physical manipulations are often obscure, at least in the sense that it is not obvious how one would construct an idealised laboratory apparatus for making any desired sort of measurement.' (*The Nature of Physical Theory* (New York: Dover, 1963), p. 118).

[60] J. M. Cook has proven that the problem in scattering theory in Hilbert space is only solvable for potentials in which $\iiint |v(xyz)| dx dy dz < \infty$ which excludes the Coulomb potential. (*Journal of Mathematical Physics*, vol. 36, 1957, pp. 82ff.) One solves this problem with the help of an appropriate cut-off. That is, one changes the potential only in order to create a connection between the theory and the facts anyway.

[61] As we see, the problems of the older quantum theory are by no means all solved. Some of them are moved to another place, and are thereby made invisible. In this respect, the older quantum theory, in which all the problems lay open and whose provisional character is clearly expressed even in its precise formalism, is much more 'honest' than its modern successor (the same goes for all versions of science which try to realize the Euclidean ideal). Here, 'honesty' is meant in Einstein's sense. See his critique of Hilbert's method of representing physical and mathematical theories to be found on a postcard to Ehrenfest from 24 May 1916, reprinted in Karl Seelig, *Albert Einstein* (Zurich, 1960), p. 276. The discrepancy between formalism and the real world which is so characteristic of the Euclidean method goes back to Parmenides. See A. Szabó, *The Origins of Euclidean Axiomatics*, London Lectures, November 1966 (MS edited by B. Burgoyne).

considerations of formal logic, like the triviality of 'Dialogical logic' [Dialoglogik], *are mainly applicable, and from which the philosophy of science develops its image of science as a consistent, logically ordered system of statements*. The proof that a system of rules or a 'logic' ('methodology') leads to interesting results in *this* domain has almost nothing to do with the question of its utility in science, that is, in its inconsistent entirety: precise theory, plus special assumptions, plus vague presuppositions, plus assignment rules, plus operational explanations, plus approximations (which quickly wipe out the arbitrary precision), plus qualitative mistakes (which quickly disappear behind the approximations), plus the theory of the apparatuses, plus 'experience' (whose nature is never explained), plus philosophical 'atmosphere'.

But even such an irrelevant proof is only seldom attempted in methodology, logic and epistemology. After all, those are special, self-contained disciplines which can and must solve their problems without outside help.[62] Thus one does not ask whether principles, rules, or methodological prescriptions are appropriate to steer the *historical process* 'science' in a certain direction. One is not in the least concerned with whether the actions demanded by the rules are psychologically, physically, historically, financially, etc., *possible*. With gay nonchalance, one compares rules with other rules, and lays the foundations of 'science', or 'knowledge', or 'reason' with that system of rules which turns out to be the winner of this abstract battle. It is as if one were so fascinated by the abstract image of a dance, that one developed and constructed it in detail without even mentioning any anatomical and physiological characteristics of the human *body*. How can we avoid this flight from reality? We will answer this question with the help of a discussion of two methodologies which played, and still play, an important role in the philosophy of science. I mean the method of verification and the method of falsification.

The method of verification (incorporate only such theories as are

---

[62] Epistemology is 'of a completely different kind than the empirical sciences [Realwissenschaften]. It does not pursue being, but sets a target and norms for mental action,' writes V. Kraft in *Erkenntnislehre* (Vienna, 1960), p. 32 [our translation]. According to this, philosophical positions, such as solipsism, are not overcome by research, but by the *decision* to use assumptions 'which lead beyond what is present in experience' (p. 219). Or: 'the study of a *largely autonomous* third world of objective knowledge is of decisive importance for epistemology' Popper, 'Epistemology Without a Knowing Subject' reprinted in his *Objective Knowledge: An Evolutionary Approach* (Oxford: Oxford University Press, 1972), p. 111. This objective knowledge 'is man-made' (Popper, 'On the Theory of the Objective Mind', reprinted in *Objective Knowledge*, p. 158), but still gives rise to a subject-less type of knowledge ('Epistemology without a Knowing Subject' pp. 108–9). Or: 'we have already seen that philosophy and historiography are at bottom irreducible to one another, no matter how closely they may be interlocked in practice', E. McMullin, 'The History and Philosophy of Science – a Taxonomy' in R. H. Steuwer (ed.), *Minnesota Studies in the Philosophy of Science, volume V* (Minneapolis: University of Minnesota Press, 1970), p. 60. See also Agassi, *Towards an Historiography of Science*, footnotes 19ff. and text.

verifiable and also already verified)[63] is subject to the *logical* objection that general statements cannot be derived from singular statements. The demand that we construct our science only out of verified statements obliterates science as we know it, without replacing it with something comparable. The method of falsification (incorporate only such theories as are falsifiable but not yet falsified)[64] has no comparable difficulties, and a science which uses it seems to be a possible enterprise. Thus far, I have provided a *purely logical* critique.

The examples in section 5 suggest the presumption that a science which obeys the principle of falsification runs into insurmountable difficulties *in our world*. Every law that we discover is surrounded by disturbances which are large enough to refute it. *In this world*, the method of falsification also obliterates science without replacing it with something comparable. Thus, a purely logical investigation of methodologies does not suffice. Which elements do we have to add to logic in order to complete our critique?

According to the view of an influential school, *scientific practice* decides the choice of methodological rules. Rules which play a role in scientific practice are accepted. Rules which cannot be found in scientific practice or which contradict the accepted rules are rejected. An 'inductive critique' of this kind overlooks that a 'practice', i.e. a series of actions, does not unambiguously determine its alleged underlying rules. And this is the case even if one is allowed to presuppose that in it, rules are never violated, overlooked, forgotten, misapplied, i.e. if one is allowed to presuppose that scientists always do the right thing with a somnambulant certainty. Practice which proceeds to use scientific theories despite contradictions with the facts could rely, for example, on the following rules: (1) it does not contain the falsification principle; (2) it contains a falsification principle, but only accepts a very special group of circumstances as the 'facts'; (3) it contains a falsification principle, but only incorporates contradicting facts if they confirm an alternative theory; (4) it contains a falsification principle, but does not apply it until a theory has had the time to develop itself; (5) the abandonment of theories is a matter of taste, and the facts do not have anything directly to do with it. And so on ... But the presupposition itself is totally incredible. The rules that we are looking for are only very rarely explicitly formulated. They do not govern practice like easily consultable books. Instead, they govern indirectly through imponderables such as tact and flair, and the interpretation of *these* is not equally accessible to every

---

[63] Details and literature in L. Laudan, 'From Testability to Meaning' (MS, London, 1968).

[64] Karl Popper writes, 'So long as a theory stands up to the severest test we can design, it is accepted; if it does not, *it is rejected.*' 'The new theory should be *independently testable.* That is to say, apart from explaining all the *explicanda* which the new theory was designed to explain, it must have new and testable consequences ...'. *Conjectures and Refutations* (New York: Basic Books, 1962), p. 54 and p. 241.

scientist. In addition, in science there are numerous traditions which overlap and are in conflict with each other in some cases. (Here science fundamentally distinguishes itself from chess, in which contradicting traditions have been melded into a unity for a long time.)[65] How else could it be explained that scientists so often thought they would find the logically impossible verification principle in their own practice? The resolute application of Rule IV relies on this belief, among other things. Moreover, it is clear that the *fact* that we follow certain rules does not *force* us, by any means, to accept the rules being followed. Although the fact that we follow rules may be the 'core of science', should we then accept science because it is already there? Isn't science itself the result of a critique of former forms of life, and doesn't this circumstance invite its own critique even more urgently? 'For after all, "the whole of science" might err ...'[66] Therefore, we cannot be content with a critique which compares methodological rules only with scientific practice. It is *impossible*, and if it were possible, it would be insufficient, because it relies on the dogmatic assumption of a certain form of life.

It is this insight that led to the development of purely abstract methodologies as early as in ancient Greece. Abstract methodologies grasp our knowledge as an ideal state which only seldom exists in our world, and which must be achieved despite diverse obstacles. The ideal is more precisely circumscribed by the rules, and an effort is made to create a practice that exactly agrees with the rules. The rules themselves are motivated differently: through the claim that they lead to the truth (by Plato), for example, or through the claim that the ascribed practice allows for an environment which promotes human liberty (by Mill).[67] As we have seen at the beginning of this section, this solution to the problem of methods is also insufficient. A purely abstract methodology may suffice logically, it may achieve a preconceived target like truth or liberty *in thought*, and it may also agree with our conception of a life worth living, but we have no guarantee that the actions described by it can be realized *in this world*.

Imre Lakatos, who noticed this last objection from a distance, tries to avoid it by a highly ingenious *synthesis* of practical and abstract considerations. He bases his proposal on two observations: (1) Although scientists argue about the *general principles* of research, their ideas about concrete

---

[65] This is overlooked in the Wittgensteinian tradition, in which one is happily content with superficial analogies.

[66] K. Popper, *Logik der Forschung* (Vienna, 1935), p. 3. [English text taken from *The Logic of Scientific Discovery* (London: Hutchinson, 1959), p. 29.]

[67] See John Stuart Mill, 'On Liberty', and the discussion of this essay in section 3 of my article 'Against Method: Outline of an Anarchistic Theory of Knowledge' in M. Radner and S. Winokur (eds.), *Minnesota Studies in the Philosophy of Science, volume IV* (Minneapolis: University of Minnesota Press, 1970).

achievements or their 'normative basic judgements', as Lakatos puts it, remained the same over the last two centuries.[68] (2) Basic judgements are not always trustworthy. They are trustworthy in physics, but suspicious in astrology and in the social sciences.[69] Furthermore, unanimity of judgement dissolves in these disciplines, so that everything becomes questionable, principles *and* basic judgements. The first observation leads Lakatos to demand that methodological rules or, more generally put, theories of rationality, are 'to be rejected if [they are] inconsistent with an accepted 'basic value judgement' of the scientific élite'.[70] The yardstick for methodological rules lies in 'common scientific wisdom'[71] which is expressed by the basic judgements. Exactly as empirical scientific theories rise above empirical basic statements, and then explain them from a general point of view, theories of rationality rise above the basic *value* judgements of the scientific élite and provide a general background for *their* explanation. The second observation makes such generalizations at least partially dependent on philosophical principles. As a result, we get a *dualism (pluralism) of authorities* in which the particular authority of basic judgements which concern only concrete individual cases both criticizes and is criticized by the general authority of philosophical (mathematical) principles. Basic judgements are fundamental in the 'mature sciences', like physics, provided that they are in a state of progress and expansion. Here the methodology is modelled on the logic of single cases. But, 'when a scientific school degenerates into pseudo-science, it may be worthwhile to force a methodological debate ...'[72] i.e., if one temporarily gives precedence to philosophical principles.

This solution to our question falters on the following difficulties.

First, the 'common scientific wisdom', i.e. the collection of all singular normative basic judgements of the scientific élite, is neither common nor wise, even in the most mature sciences. This can be shown with the help of examples, as for instance the following. According to Popper, whom Lakatos criticizes on the basis of his reconstruction of science, a theory is only scientific if the observations that could refute it are fixed in advance,[73] and it must be abandoned as soon as one discovers the first refuting instance.[74] On the other hand, Newton's theory 'is highly regarded by the

---

[68] 'History and its Rational Reconstructions' (MS, Boston, 1970), p. 31ff. [English citations taken from 'History of Science and its Rational Reconstructions', reprinted in I. Lakatos, *The Methodology of Scientific Research Programmes: Philosophical Papers, volume 1*, J. Worrall and G. Currie (eds.) (Cambridge: Cambridge University Press, 1978). Page numbers for text taken from this source are given in square brackets.]

[69] *Ibid.*, p. 51 [p. 137, note 4].

[70] *Ibid.*, p. 31 [p. 124], see also pp. 32, 36, 39, 43, 46, 50.

[71] *Ibid.*, p. 50 [p. 137].   [72] *Ibid.*, p. 51 [p. 137].

[73] *Conjectures and Refutations*, p. 38, note 3.

[74] *Conjectures and Refutations*, p. 54 and p. 241. See also my citation in footnote 64 of this article.

greatest scientists',[75] even though it contradicted numerous facts throughout its existence, and even though a clear definition of its empirical content (in the Popperian sense) never existed. 'Newton's theory is scientific' is therefore a normative basic judgement in Lakatos' sense. Therefore Lakatos concludes that Popper's definition of science is refuted.[76] But let us pay attention to how these basic judgements on which he relies came about in this specific case. Max Born[77] praises Newton's mechanics in the belief that it logically follows from observable facts (in the nineteenth century, this opinion can be found more often. Exceptions are Hegel and Duhem.[78]). The majority of Newtonians support the theory because it is without a blemish according to their view. Newton himself claims that he derived his theory directly from 'phenomena', without even glancing at the alternatives. *Those* are the reasons which lead to the basic judgement of 'scientific wisdom' to which Lakatos gives so much confidence. And let us not forget that Einstein's theory was rejected by a large number of physicists in the years 1907 and 1908 because it contradicted the facts.[79] Thus, basic judgements are not as general as Lakatos likes to assume. And that is a result which can be confirmed repeatedly by further investigations. The idea of 'common scientific wisdom', i.e., the idea of a comprehensive class of singular value judgements which are *reasonable* and *generally* accepted in science, this idea is nothing other than a *chimera*. With that, the 'dualism of authorities' dissolves into nothing.

Let us now assume that there really is a solid core of normative basic judgements. Then it still remains undecided when this core, or when its opponent in dualism, i.e. when philosophical principles, should take centre stage. Lakatos pushes philosophical principles into centre stage as soon as the underlying science 'degenerates into pseudo-science' (see text to footnote 72 above). But how does he decide what is and what is not a pseudo-science? He makes the decision on the grounds of a 'rational reconstruction of science',[80] which takes normative basic judgements as its starting point. This presupposes that these basic judgements possess a certain uniformity which, according to his own opinion, is precisely not the case when philosophical principles take centre-stage. Thus the criterion for pseudo-science that Lakatos uses already presupposes what should be shown, i.e., that good science produces uniform basic judgements.

Third, Lakatos is also exposed to the objections raised above against the criticism on the grounds of scientific *practice*. Although Lakatos does not try to gain *general* rules from practice – and such an undertaking, according to

---

[75] Lakatos, *Methodology*, p. 34 [p. 125].   [76] *Ibid.*
[77] *Natural Philosophy of Cause and Chance* (Oxford: Clarendon Press, 1949), pp. 129ff.
[78] For Hegel, see the *Encyclopädie* (ed. Lasson) pp. 235ff.
[79] See footnotes 32–7 of this article, and the corresponding text.
[80] Lakatos, p. 41 [p. 127].

his opinion, does not lead to a uniform result – he restricts himself to extracting *singular* evaluations, just his 'normative basic judgements'. A criticism which uses *these kinds* of means assumes that science has already found the best possible procedure, at least in concrete cases. That is childish optimism. Science has seen methodological revolutions side by side with revolutions in the content of its theories. The transition from the Aristotelian to the Galilean approach is an example. Such revolutions overthrow not only one point of view or another, but *all* ideas which were founded on the grounds of certain procedures, basic judgements included. A critique which leaves basic judgements untouched is therefore much too tame. And if *science*, in the concrete, is so perfect, why isn't myth so perfect? What are the circumstances which distinguish modern science, or is its perfection only a matter of faith which is no longer allowed to be questioned? We see that Lakatos' criticism of methodology, let alone the falsity of its presuppositions, is satisfied much too soon. Let us now use our example (verificationism vs. falsificationism) in order to explore the path to a more reasonable philosophy!

The method of falsification is better than the method of verification. An argument on the grounds of normative basic judgements would refer to the fact that scientists shower compliments on unverified theories, and that the critique of theories is more important to them than their proof. This argument is unsatisfactory. The judgements from which it starts are often falsely justified. They are the result of large mistakes, and are totally unsuitable as a basis for methodological conclusions. Newtonian theory, which contradicts numerous facts *de facto*, is praised because it is believed that it follows logically from the facts. Other theories, for example the general theory of relativity, are criticized because they are too distant from experience. A better argument against verificationism consists in the claim that verification is logically impossible, while falsification is logically possible. *This* argument draws on certain general relationships between states of affairs. It is a *logical* argument.

The method of falsification is better than the method of verification. But it is still vulnerable to criticism. Following Lakatos, one can say that 'respectable' theories (a normative basic judgement!) are almost always in contradiction with clear experimental facts. This argument is also unsatisfactory. Quantum theory, which has qualitative and quantitative difficulties, is accepted not *despite* these difficulties, but because 'all the evidence proves with merciless definiteness the conformity of the unknown interactions with the fundamental quantum law'.[81] And where the discrepancy between facts and theory penetrates the threshold of the scientist's consciousness, the

---

[81] See footnote 45 and text [and accompanying translator's note].

theory is actually rejected by the vast majority of scientists.[82] We see again how *unreasonable* and *inconsistent* normative basic judgements are, and how little the attitude of the 'scientific élite' is suited as a foundation for methodological discussions. A better argument consists *here* in the remark that the laws of the world in which we live are surrounded by numerous disturbances and are even partially disguised. The method of falsification (which was an important element of Aristotle's method[83]) promotes science as long as such disturbances disappear behind experimental mistakes, or as long as one considers them to be monstrosities which do not stand in any relationship to underlying laws.[84] One gets the impression, which is characteristic of Aristotelian philosophy, that the laws of nature lie open to us, and that obstinate observations are correctly conceived of as a proof that our *theories*, but not the *methods* used by us, are fraught with mistakes. A problem arises as soon as the disturbances become a daily affair, which is also indeed the case in modern science (see the small collection of them in Section 5). We are now confronted with the following alternatives: (a) We retain the method of falsification, and conclude that 'knowledge', in the sense of this method, is not possible in our world.[85] (b) We change our idea of knowledge. We replace it with a more abstract (less empirical, less critical) idea, and choose a more liberal methodology as its basis. Many scientists instinctively act as if they have consciously considered the difficulty just described, and then chose to accept alternative (b). An investigation of their reasons shows, as already intimated, that conscious analysis is out of the question. It is a combination of errors, fallacies, prejudices, greed, sheer stubbornness, in other words, it is not reason, but the *cunning of reason* which leads the scientists in this direction *sometimes*. Therefore you cannot rely on 'basic judgements'. One must replace them with both *cosmological hypotheses*, like for instance with the hypothesis that laws of nature are not manifest in our world, and with the selection of rules which do not fail in a world described by such hypotheses.

---

[82] See footnotes 32–7 and text.   [83] See footnote 24.

[84] See Kurt Lewin 'Der Übergang von der Aristotelischen zur Galileischen Denkweise in Biologie und Psychologie', *Erkenntnis*, vol. 1, 1931.

[85] This conclusion was already reached by the authors of the Hippocratic text 'The Ancient Medicine'. Natural philosophers proposed general principles, while medical practice showed that every substance only had a certain effect, under certain circumstances, at certain times, with certain people, etc. Therefore the treatment of concrete cases had to take the place of building theories. This attitude was strengthened in the Hellenistic Age by the influence of scepticism. See L. Edelstein, 'Empirie und Skepsis in der Lehre der griechischen Empirikerschule' in *Quellen und Studien zur Geschichte der Naturwissenschaften und der Medizin, 3, 4* (Berlin, 1933), pp. 45ff. In the Hellenistic Age, this argument then led to a new interpretation of experience (footnote 24). A third possibility consists in not noticing this problem, and pushing aside disparaged theories with the claim that they are in obvious contradiction with the facts. The critique of Marxism by Bernstein and Popper belongs in this category.

This *cosmological* critique of methodological rules is content neither with an *abstract* comparison (of rules with other rules), nor with a comparison of method and *practice*, nor with a *combination* of both procedures. Science should proceed in a particular world, under particular historical, psychological, and physical conditions. Therefore these conditions must be taken into account. They are taken into account by proposing cosmological hypotheses, and then suggesting rules which can lead to interesting, rich, and ... results in a world described by these hypotheses, in which '...' expresses any additional demands. The first cosmological hypothesis with which one begins the procedure is, of course, found and justified in the usual way (for exceptions see below). That is, for its proposal, one draws from the customary methodological procedure. The procedure is criticized and replaced by other procedures which lead to further hypotheses, until a methodology is finally found which justifies a certain cosmological hypothesis, and is also recommended by it. Schematically:

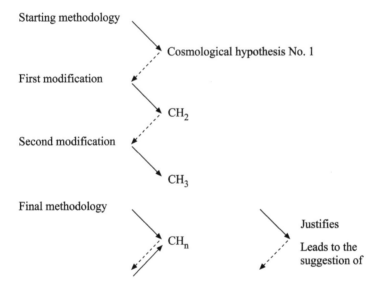

We now come to the question of where one finds the *evidence* for a cosmological hypothesis. The claim that, after all, one only needs to make observations and experiments which then lead to hypotheses in the usual way does not suffice. It does not suffice because the 'usual way', that is the accepted methodology, not only decides questions about hypotheses, but also decides which circumstances can enter the frame of science as 'facts'. Circumstances which do not obey this methodology are pushed aside as errors, superstitious illusions, dreams, mythologies. The honorary title 'fact' is not awarded to them. Even methodologically permissible observations which run counter to a popular theory seldom enter the representation of the situation of this theory, and their scientific relevance is often completely denied.[86] But exactly such observations, exactly such pseudo-facts, are of the highest relevance for our critique. They are evidence for cosmological hypotheses which suggest other standards, which include other distinctions between facts and pseudo-facts. Where do we find the reports which contain them?

Surely, we do not find them in official representations of our knowledge, which are already chosen and composed from a certain point of view. Nor do they appear in the official history of the sciences, because they try to disguise all errors and all evidence which should never have appeared according to contemporary theories. We find the desired reports in the more obscure corners and niches of *general history*, the history of ideas, the history of superstition, of errors, of delusion. Also, older scientific journals, which have already been forgotten and were not subjected to the tyranny of a certain method, contain a wealth of 'facts' of the kind we need.

---

[86] So, for example, for a long time facts which seemed directly to support the idea of possession by the devil or the idea of demonic influences were suppressed, and this delayed the development of humanistic psychology for centuries. The authors of the *Malleus Maleficarum*, writes Gregory Zilboorg, possess 'a rather complete knowledge of the symptomatology of mental diseases which they use exclusively for purposes of detecting witches. They knew of the hysterical anesthesias which they carefully investigated by means of torture, and through the same means they elicited the phenomenon of extreme pathological mutism, which is so frequently found among stuporous catatonic schizophrenias. They speak of various contortions of the body – evidently hysterical convulsions... Sprenger and Kraemer described literally every single type of neurosis or psychosis which we find today in our daily psychiatric work.' 'The *Malleus Maleficarum* might with a little editing serve as an excellent modern textbook of descriptive clinical psychiatry of the fifteenth century, if the word *witch* were substituted by the word *patient*, and the devil eliminated.' At that time, 'the whole field of clinical psychiatry was covered by theologians ...' (*The Medical Man and the Witch During the Renaissance* (Baltimore, 1935), pp. 49ff., p. 58, p. 78). This circumstance by itself was enough to treat not only the theories, but also the factual reports of the *Malleus* as mere dreams for a fairly long time. O. Temkin reports a similar development in other areas of medicine, *The Falling Sickness* (Baltimore, 1945), pp. 225ff. In astronomy and in terrestrial physics, it was no different. Connections between astronomy and biology were studied in former times by astrology. As Comte writes, 'these connecting links between astronomy and biology were studied from a very different point of view, *but at least they were studied and not left out of sight*, as is the common tendency in our time, under the restricting influence of a nascent and incomplete positivism. Beneath the chimerical belief of the old philosophy in the physiological influence of the stars, there lay a strong though confused recognition of the truth that the facts of life were in some way dependent on the solar system. Like all primitive inspirations of man's intelligence this feeling needed rectification by positive science, but not destruction; though unhappily in science, as in politics, it is often hard to reorganize without some brief period of overthrow' *Philosophie Positive* (Paris: Littré, 1836), III, pp. 273ff. [English taken from Feyerabend's translation from the French in his article 'Against Method: Outline of an Anarchistic Theory of Knowledge', p. 125, note 184.] See footnotes 20, 21, and text, as well as Galileo's theory of the tides, which can be traced back to his refusal to acknowledge heavenly influences on earthly events, *Dialogue* (ed. Drake) pp. 419ff.

Superficially seen, the cosmological critique resembles a critique on the grounds of scientific practice. Both begin with history. But while the *one* looks for rules in history which the practice *de facto* obeys, and uses the found rules as cogent arguments in methodological debates, the other is mainly interested in the circumstances, facts, pseudo-facts, etc., which can be considered as the starting points for cosmological generalizations. The generalization serves, then, as the criterion for the realizability of instructions, rules and methodological systems which themselves are obtained on the grounds of *decisions*. Here, we are interested in the individual case, which, as we know, could be interpreted differently or could be changed, while the others investigate what is *possible* in our world and reject impossible demands. The cosmological analysis of methodologies does not make the mistake of believing that the mere *existence* of a practice already suffices for the critique of philosophical principles. Against the remark that scientists use refuted theories without batting an eyelid, it answers (together with the thinkers of the abstract school) that the science of today is surely not the last word, and that unreasonableness does not become reasonable only because it has a battalion of Nobel prize laureates standing behind it. But it is also not content with the philosophical plausibility or the logical excellence of a rule. Quite to the contrary, it would consider it to be a serious objection if one were to show that abstractly described and logically praiseworthy procedures not only do not appear in practice, but also *cannot appear* for (physical, physiological, historical, etc.) reasons. This kind of critique also begins with history, but it does not ask what actually happened, but instead which sequences of actions have the prospect of succeeding, and which are damned to failure. We see that the role of history is entirely different in the two cases.

The dissimilarity must be emphasized especially in those cases in which we use *historical generalizations* side by side with cosmological generalizations in our methodological arguments. After all, the attempt to increase our knowledge does not depend only on physics. To perhaps an even greater extent, it depends on the historical circumstances (ideologies, institutions, scientific and other stimulations) in which problems are posed, analysed, and solved. It also depends on the laws which such circumstances obey. These laws can also prevent the realization of demands, and they are an additional element that we must take into consideration in the critique of methodologies. They are often quite trivial and surely not valid without exceptions, and it might perhaps be better to replace the word 'law' with another word. On the other hand, there are *tendencies* here, which are not simply subject to the will of the individual. They confront him as an objective resistance (example: the tenacity of traditions), and make themselves felt not only in a certain age, but also in distant stages of the development of the sciences. The word 'law', if not taken too seriously, is indeed therefore justified to a certain extent. Examples explain this point better than abstract considerations.

(1) Knowledge of observation instruments (eyes, glasses, magnifying glasses, telescopes), indeed even the mere interest in these things, is often less developed than the knowledge of the world that we investigate with their help. Physics, astronomy, etc., and physiology are almost never 'in phase'. The last trails far behind the first.[87]

(2) The physiological assumptions that one takes to be true in certain periods are only rarely explicitly formulated. The vast majority enter the observation language. They constitute the rules of the language and thereby the content of the observation concepts as well.

(3) Theories become promoted especially by ideas which agree with them (this seems to be very trivial, but leads to an important argument against empiricist philosophy of science, see below). More especially, a modern physiological (psychological) theory flourishes more easily in the presence of modern astronomy, than in the presence of out-dated ideas. Physiology discovers the relationship between objective stimulus and subjective sensation, and it will progress more quickly if an adequate representation of an objective stimulus is already at hand, or at least if one has a choice between different representations of this stimulus.

These three tendencies, or 'laws', have the following effect on the structure of our knowledge.

First, the *disturbance* of the process of cognition by events in our

---

[87] This is a special case of a much more general regularity which Marx has already captured and named 'uneven historical development'. See *Das Elend der Philosophie und besonders die Einführung zur Kritik der politischen Ökonomie* (Berlin: Dietz Verlag, 1963), p. 257. Trotsky describes the same situation: 'The gist of the matter lies in this, that the different aspects of the historical process – economics, politics, the growth of the working class – do not develop simultaneously along parallel lines.' [English taken from Feyerabend's 'Against Method: Outline of an Anarchistic Theory of Knowledge' p. 118, note 127, which cites *The First Four Years of the Communist International, volume II* (New York: Pioneer Publishers, 1923), p. 5.] See also Lenin concerning the fact that different causes of an event are not always in phase, and in such cases, lead to no effect, *'Left Wing' Communism, an Infantile Disorder* (Beijing: Foreign Language Press, 1965), p. 59. In another form, the thesis of 'uneven development' is concerned with the fact that capitalism in different countries and in different parts of the same country is unequally developed. This can lead to an inverse relation between the accompanying ideologies: 'In civilised Europe, with its highly developed machine industry, its rich, multiform culture, and its constitutions, a point of history has been reached when the commanding bourgeoisie, fearing the growth and increasing strength of the proletariat, comes out in support of everything backward, moribund, and medieval ... But all young Asia grows a mighty democratic movement, spreading and gaining in strength' ('Backward Europe and Advanced Asia', *Collected Works, volume 19*, pp. 99ff.). For these very interesting situations that deserve to be more closely observed by the philosophy of science, see also A. G. Meyer, *Leninism* (Cambridge, 1957), ch. 12, as well as L. Althusser, *For Marx* (New York: Vintage Books, 1970), chapters 3 and 6 (though Althusser's philosophy of science is stuck in the darkest Middle Ages of thought). On the whole issue, also see Mao Tse-Tung's essay, 'On Contradiction', especially section IV.

instruments and sense-organs (cerebrum and cerebellum included) is inaccessible to the grasp of science, even given a highly developed astronomy and physics. It is only seldom that there are useful explanations for the deviations that result from them, and these deviations are also not allowed to be played off against new theories, even if they surpass the limits of measurement precision.

Second, the *descriptions* of the observable facts contradict a theory often only because the concepts with which they were formulated belong to older theories. In this case, the contradiction is not between theory and 'fact', but between newer theory and older theory. However, the older theory does not openly expose itself, but disguises itself in apparently totally harmless observation reports.

Third, for the reasons just mentioned, and also because of 'law' number 3, it is often appropriate to replace seemingly fundamental presuppositions which were held as certain for a long time, such as the presupposition of the relative permanence of our measuring instruments and our experimental records, with unproved conjectures which harmonize the observation material with new and doubtful theories.

The discrepancy between the Copernican theory and the results of direct observation, which were sketched in the text to footnotes 54 and 55, formidably illustrates these considerations. Let us briefly consider the objection on the grounds of the brightness of Mars and Venus.[88] The *observed* brightness of Mars and Venus changes much less in the course of a year than one would assume on the basis of Copernican calculations. For naïve realism, which assumes an exact correspondence between external stimulus and optical impression, this means the falsity of the Copernican theory. This was also the position of the Aristotelians. As we know, they had a detailed theory of perception which was in accordance with their physics, and which was completely confirmed by simple experiences.[89] On the other hand, one can presume that the human eye does not always represent the environment correctly. As Galileo puts it, 'rather, the very instrument of seeing introduces a hindrance of its own'.[90] Galileo does not have a theory of these hindrances. This only puts him at a disadvantage when compared with the Aristotelians if one forgets the phase difference mentioned in (1). If one takes these phase differences into account, then the following possibilities arise:

(1) We assume that the problem of the brightness of Mars and Venus is a case which restricts the validity of naïve realism, even in its highly

---

[88] For a more detailed treatment which also considers the role of the telescope, see my article 'Problems of Empiricism, Part II' in R. G. Colodny (ed.), *The Nature and Function of Scientific Theories* (Pittsburgh: Pittsburgh University Press, 1970), where the role of 'laws' 1–3 is extensively discussed.

[89] *De Anima II*, v–vii, xii.   [90] *Dialogue* (ed. Drake), p. 335.

developed Aristotelian form, and we *abstain from a judgement* until a more thorough investigation of sense organs leads to an alternative theory. The Copernican theory of the heavenly bodies was preceded by the Aristotelian theory of perception. Thus one has to *wait* until physiology catches up to astronomy.

(2) One does *not* abstain from a judgement, but declares the still completely hypothetical Copernican theory *true*. We expand it and use it as an auxiliary for detecting the desired neuro-physiological principles. Naïve realism, on the other hand, is removed from consideration despite its massive foundation. This is the procedure that Galileo chooses, although the propaganda with the telescope does not make it easy for us to discover the features of this procedure.[91] With this he shows astonishing foresight. The phenomenon of *irradiation*, which is partially responsible for the discrepancy, is not completely understood even today. (Just look for it in Davson's four volume Opus, *The Eye!*)

Considerations like this show that 'the orthodox view on the nature of theories',[92] which is widespread today, has very little to do with science. According to this view, our knowledge consists of layers of different certainty, so that one gradually ascends from facts to certain laws, to highly confirmed theories, to highly doubtful hypotheses. The investigation of doubtful hypotheses draws from different auxiliary assumptions, and those 'have usually been "secured" by previous confirmation'. Feigl continues, 'it would be foolish to call them into doubt when some other more "risky" hypotheses are under critical scrutiny'.[93] The erection of the layer system, 'knowledge', thus begins from beneath and slowly progresses upwards. Criticism is directed at the upper regions, and it leaves the 'basis' relatively untouched, at least as long as one compares ideas in those regions, '. . . pervasive presuppositions, for example, regarding the relative permanence of the laboratory instruments, of the experimental records, are "theoretical" [i.e., not totally certain and formulated in abstract concepts] only from a deep epistemological point of view and are not called into question when, for example, we try to decide experimentally between rival theories in the physical, biological, or social sciences'.[94] Contrary to this, our little analysis of the situation in Galileo's time shows that a 'pervasive presupposition' (like naïve realism, which is ' "secured" by previous confirmation' and is 'theoretical', not in any practical sense, but in a highly ethereal sense of a super-critical epistemology) is in fact not only profitably doubted, but even profitably rejected while one continues to retain the highly 'risky' hypothesis of the motion of the earth despite observations to the contrary (one

---

[91] For details, again see 'Problems of Empiricism, Part II'.
[92] This is the title of the leading article from Feigl in *Minnesota Studies in the Philosophy of Science, volume IV.* [The actual title is 'The Orthodox View of Theories'.]
[93] Feigl, *ibid.*, p. 10.   [94] *Ibid.*, p. 13.

must consider the situation in the late sixteenth and the early seventeenth century correctly). It is profitable to investigate and describe somewhat more exactly this new interpretation of knowledge which now has to replace the layer theory.

7 KNOWLEDGE AS A HISTORICAL PROCESS

Methodology and epistemology have almost always looked at scientific problems *sub specie aeternitatis*. One compares statements with other statements without considering their historical development, and without taking the fact that they may belong to different historical periods into consideration. So, for example, one poses the question: how do observations, initial conditions, basic principles, and philosophical considerations shed light on a particular theory? One gets very different answers to this question. According to one school, it is possible to calculate degrees of confirmation from given material, and to judge the theory with its help. Others reject any kind of logic of confirmation, and compare theories according to their content and the number of known refutations. Other thinkers cherish the illusion that theories can be derived logically from observations. But to all of these schools, it is self-evident that exact observations and clear principles already decide the fate of a theory, that one has to use them here and now either in order to reject a proposed theory, or to confirm it, or even to prove it.

Such a procedure is only meaningful if one can assume that the elements of our knowledge: our theories, our observations, the principles of our arguments, are all *timeless entities* which are all equally perfect, equally accessible, and all stand in particular relationships to each other, no matter what events led to their discovery. This assumption is widespread. It has its source in the well-known distinction between the context of discovery and the context of justification. It is often expressed in the claim that science is occupied with propositions, but not with statements or sentences. This, however, overlooks the fact that science should be considered as a complex and a highly heterogeneous *historical process* in which vague and unrelated anticipations of future ideologies are developed side by side with highly sophisticated theoretical systems and petrified forms of thinking. Certain elements of this process are neatly recorded (published) in compendia, while others have an underground existence which can only be discovered by contrast, that is, by comparison with new and extraordinary ideas. Numerous conflicts and contradictions have their roots in the heterogeneity of the material and are of no theoretical importance. They resemble the problems which arise when a high-voltage power station is needed in the vicinity of a gothic cathedral. This unevenness of science is sometimes noticed, as in the claim that physical laws and biological laws belong to different conceptual systems and simply cannot be compared with each other. But in most cases, and especially in the comparison between observation and theory, our methodologies project all elements of science, and the different historical layers to which they belong, on to one and the same level, and then compare these caricatures on this level. This is as reasonable as the attempt to arrange a boxing match between a toddler and an adult athlete, together with a triumphant announcement that the adult will surely win. (The history of the classical atomic theory and the much shorter history of hidden variables in quantum theory is full of idiotic remarks of this kind. The same applies to the history of psychoanalysis and of Marxism.) It is clear that when we investigate newer theoretical proposals, we have to take the historical situation into consideration. Let us see how this will influence our judgement!

The geocentric hypothesis and the Aristotelian philosophy of science are ideally matched. The theory of motion which has the stationary earth as a consequence is supported by perception, and it is itself a special case of a more comprehensive theory of motion which deals with local motion, increase and decrease, qualitative change, coming to be and passing away, and other special events. According to this comprehensive theory, motion, and indeed any kind of change, consists in the transition of a form from the cause to the influenced body, and it stops as soon as the latter possesses exactly the same form which the cause featured at the beginning of the process. According to this theory, perception is also a process in which the form of the perceived object enters the perceiving organ via the detour of a medium. Then the form in the perceiving organ is once again the same as the form of the object perceived, so that the perceiver, in a certain sense, adopts the properties of the perceived object.[95]

A physiology of this kind, which is actually nothing other than a highly developed version of naïve realism (see section 6), leaves no room for a discrepancy between observation and reality. 'The tradition's concept of nature was connected with a kind of *visibility postulate* which corresponded both to the finiteness of the universe as well as to the idea of the utility for and the centrality of human beings in the concept of nature. That there should be things in the world which are inaccessible to man not only now, and for the time being, but in principle, and because of his natural endowment, and which would therefore never be seen by him – this was quite inconceivable for later antiquity as well as for the Middle Ages.'[96] The theory also does not allow for the use of *instruments* because instruments

---

[95] 'Moreover that which sees does in a sense possess colour'. *De Anima* 425b24. [English translation of Aristotle taken from *On the Soul. Parva Naturalia. On Breath.* Translated by W. S. Hett (Cambridge, MA: Harvard University Press, and London: Heinemann, 1952), p. 147.]

[96] Hans Blumenberg (ed.) Galileo Galilei, *Sidereus Nuncius, Nachricht von neuen Sternen* (Sammlung Insel, 1965), p. 13. [This is in Blumenberg's introduction, not in Galileo's text.]

(telescopes, microscopes, etc.) distort the process in the medium which is responsible for an exact transfer of the forms. We receive forms which no longer agree with the figure of the perceived object, so-called *illusions*. Such illusions are indeed seen in a concave mirror or in glasses with round external surfaces in which one sees pictures with coloured edges, distorted contours, blurred details inaccurately localized.[97] In Aristotle, astronomy, physics, psychology, and philosophy all work together to create a system which is coherent, comprehensive, rational, and empirically adequate.[98]

   This system and the evidence which supports it was flatly denied by the successors of Copernicus. According to the view of the Copernicans, there are cosmic processes of fabulous extent *which leave no trace in our experience*. Therefore, the existing observations are no longer considered to be tests of the new laws being proposed. They are not directly connected with these laws, and it is possible that they are in no relationship at all to interesting astronomical events. *Today, after* the success of Copernican science taught us that the relationship between the human being and his surrounding world is much more complex than Aristotle assumed, we are in a position to admit that the Copernicans had indeed made a correct guess. The observer and the basic laws are separated from each other by (1) the special physical conditions of the observation platform, i.e., the earth quickly rushing through space (effects of earthly gravity, the law of inertia, coriolis forces, atmospheric influences such as refraction, and so on) (2) the idiosyncrasies of the instruments used for observation, such as the human eye (irradiation, after-images, lateral inhibition of neighbouring retinal elements, and so on), as well as (3) older ideas which penetrated the observation language and make this language an automatic preacher of naïve realism (compare the remarks in section 6, point 2). Of course observations can contain a contribution from the observed object, but this contribution is usually overlaid by other effects, and sometimes even totally wiped out. In order to understand this, look at a picture of a fixed star as we see it through a telescope. First of all, the image is displaced by refraction, aberration, and gravitational effects. It contains a spectrum of the star, not in its contemporary form, but from a time long past (in the case of extra-galactic novae, it could be millions of years) which is distorted

---

[97] For the role of these illusions in the arguments about Aristotelian philosophy, see V. Ronchi's article 'Complexities, Advances and Misconceptions in the Development of the Science of Vision: What is Being Discovered?' in A. C. Crombie (ed.), *Scientific Change* (London: Heinemann, 1963).

[98] Like every other 'system', naturally, Aristotle's is also full of lacunae, contradictions, bad excuses, and the later writings of the scholarly author are not always in accordance with his earlier writings. See I. Düring, *Aristoteles* (Heidelberg, 1966). *Potentially*, however, the system is much more uniform than any other that followed. When we talk about Aristotle, we always mean this potential system (whose parts were accomplished through his successors to a larger and larger extent) and *not* the *corpus Aristotelicum*.

by the Doppler effect, galactic matter, and so on. The extension and the internal structure of the image depends *totally* on the telescope and the eye of the observer. It is the telescope which determines the size of the diffraction disclets, and it is the human eye which determines how much of this structure enters the consciousness of the observer. It requires training *and a lot of theoretical assumptions* to isolate the contribution of the source of the image as it is finally perceived, and to prepare it for a test. But this means that one can test non-Aristotelian cosmologies only after one has connected observations and laws by *auxiliary sciences* which describe the complex processes between the eye and the object, and the even more recondite processes between the cornea and the brain. In the Copernican case, we need a new *meteorology* (in the good old sense of the word: a science of the things happening between the moon and the earth's surface), a *physiological optics*, a new *dynamics*, and so on. Observations only become relevant *after* the processes between the eye and the world, as described by these sciences, are inserted. The language in which we describe our observations must also be carefully investigated, so that the new cosmology is not betrayed by a disguised cooperation of sensations and older ideas. Thus, *testing the Copernican theory presupposes a new world view, with a new perspective on human beings and their capacities to know.*

Now it is clear, and here we only repeat what was already said in the last section about 'phase differences' between physics and psychology, that the erection of such a new and complex world view cannot happen overnight. It is very improbable that the idea of the motion of the earth should produce all the required auxiliary sciences at the same time. Today Copernicus – tomorrow Helmholtz: this is not just improbable, but also impossible in principle, given even a brief consideration of the nature of man, his surrounding society, and the complexity of the physical world. And still a test is only meaningful when all the described sciences are already clearly and simply formulated.

Therefore, we must first *wait* and *ignore* many observations. Up until now, not a single empiricist has taken this into account. Without even suspecting the need for new types of observation and new types of criteria, empiricists immediately bring new theories together with the status quo and announce triumphantly the triviality: 'the theory contradicts the facts and the accepted principles'. They are of course right, but not in the sense intended by them. In the early stage of a new theory, a contradiction shows nothing other than that it is *new* and *different* from older theories, concepts, and observations. With this, a value judgement is not yet made. Such a value judgement presupposes that the opponents are standing on level ground. How are we to proceed in order to accomplish such a fair comparison? The first step is clear: we have to retain the new cosmology until it is complemented by the required auxiliary sciences. We must retain it in the face of clear and unambiguous conflicting observations. We can, of course, try to explain our action by claiming that critical observations are irrelevant or illusory, but we do not have an objective reason for such an explanation. The explanation is nothing but a *verbal gesture*, a polite invitation actively to participate in the development of the new cosmology. It is still possible for us to refute the accepted theory of perception. This theory declares the existing observations relevant, and gives reasons for the claim that they were confirmed by independent evidence (see above, on the Aristotelian theory of perception). The new theory, therefore, is totally cut-off, on purpose, from the data which supported its predecessor. One intentionally increases its 'metaphysical' character. This is how a new period in the history of science begins, from an empirical point of view, with a *step backwards*. We return to an earlier stage when theories were more vague and had a smaller empirical content. This step back is not a mere coincidence, but instead has a well-defined function. It is essential if we want to overtake the *status quo*, because it gives us the time and the liberty which is needed to develop new ideas in detail and to find the necessary auxiliary sciences.

(The 'step back' is therefore really a step forward. Furthermore, because every idea which contradicts a given experience and its principles can become the centre of a new cosmology, it is always an advantage if one does not take experience and experiment too seriously. An exception exists only if the senses or the apparatus themselves are doubted on the grounds of fixed ideologies. In this case, the argument is reversed and it is now experience which is used for attaining progress.)

The 'step back' is therefore really essential, but how can we persuade our opponent to follow our proposals? How can we seduce him out of his well-defined, complicated, and empirically highly successful system, in order to interest him in an unfinished and absurd hypothesis? In a hypothesis which is contradicted by one observation after another, when one only needs to make the effort to compare it with the impressions of our senses and the results of our experiments? How can we persuade him that the success of the *status quo* is only apparent, and that proofs of this will turn up in the next five hundred years, if we, *here and now*, cannot muster a single argument on our side? (And one should consider that the illustrations used above derive their power from the success of classical physics, and that this success was not available to the Copernicans. *They* could only enlist the philosophies of Heraclitus, Democritus, and the Sceptics in their favour.) Reason, 'scientific method', does not help us further here. We must employ 'irrational methods'. We need these methods in order to maintain the conviction, which is nothing else but blind faith, until the means which allow us to transform this blind faith into crystal clear insight are available.

It is this context which makes the rise of a new secular class with a new ideology, and a contempt for the 'school wisdom' so terribly important: the barbarian Latin of the scholars (which has much in common with the no less barbarian 'ordinary English' of the Oxford 'philosophers'), the intellectual poverty of academic science, their out-of-touchness, which one can also interpret as uselessness, their relationship with the church – all of these elements are thrown together with Aristotelian cosmology, and the contempt one feels for them transfers itself to every Aristotelian argument.[99] This 'guilt by association' does not make the arguments less 'rational', but it reduces their influence on the thoughts of those who are ready to follow the ideas of Copernicus. Because Copernicus stands for progress also in other fields, he is a symbol for the ideal of a new class which looks back to the Classical period of Plato and Cicero, and forward to a more free and open society. Thus the alliance of astronomy, historical tendencies, and class tendencies does not produce new arguments. Nor does it determine the form of the laws which are to be discovered (this last assumption can often be found in over-enthusiastic, and somewhat superficial interpreters of Marxism on the relationship between 'bases and super-structures'). But this alliance leads to a very strong belief in Copernicus, and this is all that is required. Galileo exploits the confusion with mastery, and expands on it with his own tricks, jokes, and non-sequiturs.

This is the situation that we must analyse and understand if we want to put the dispute between 'reason' and 'irrationality', between 'method' and 'anarchism in thought', in its true light. Reason, which after all always appears in combination with a certain method, admits that the ideas which we introduce in order to expand and increase our knowledge may possibly *arise* in a very chaotic way, and that the historical source of a cosmology may very well depend on class prejudices, passions, personal idiosyncrasies, and questions of style. But it demands that when we *judge* such ideas, we strictly follow the rules of a certain methodology (this is what the distinction between 'the context of discovery' and 'the context of justification', so often discussed in modern philosophy of science, boils down to). But our historical example (which is generalizable in the way described in section 6) shows the following: there are situations in which even the most liberal methodology, and even the most liberal conception of the laws of reason, would dispose of an idea which later plays a large role in science. The idea was proposed. It survived, and now it is in the centre of astronomy. It survived because prejudices, passions, delusion, illusion, errors, dull stubbornness, in short all of those elements which characterize the 'context of discovery', *resisted* the dictates of reason, and it survived because these

---

[99] For this situation, see L. Olschki's excellent work, *Geschichte der neusprachlichen Wissenschaften Literatur*, 3 (Reprinted Vaduz, 1965). See also R. F. Jones, *Ancients and Moderns* (Berkeley: University of California Press, 1965), chapters v and vi.

irrational elements *predominated* in the end. Put differently, *Copernicanism and other 'rational' ideas only exist today because 'reason' was often out-voted in the past.* (The inverse is also true. The belief in the devil, in the witch craze, and other 'irrational' ideas have lost their influence because reason, in the course of *its* history, was often out-voted.)[100]

We can now safely assume that the Copernican theory is scientifically flawless. Therefore, it was advantageous that it survived until today. Therefore, it was advantageous that 'reason' was out-voted in the sixteenth and early seventeenth centuries. The cosmologists of these centuries did not have the knowledge that we have today. They did not know that the teachings of Copernicus would lead to a flawless system. They did not know which of the many ideas existing at their time would lead to rational results through an irrational defence. They could only guess without arguments, only follow their inclinations. But in this respect, our situation is exactly the same. Therefore, it is also sensible today, despite all methods, to follow our own inclinations, and to claim that science will make use of such a procedure someday. This is because science must prove itself in the rich, complex surroundings of the history of nature, as well as of society, and not in the sealed-off study of the methodologists. The rules of the latter not only do not suffice for methodologies to function, but also shackle the methodologist, seriously endangering future development, and with that, the development of our consciousness.

8 THE ROLE OF METHODOLOGY

What can we expect given this methodological situation? What are the rules that we must follow in order to arrive at useful results in science as well as in practice (like for example in politics)? Which steps lead to success and which are to be avoided? After the considerations just adduced, the answer is clear. Although rules which are valid in all possible circumstances can be set up and even possibly violently enforced, it will always be *at the expense of the possibility of fundamental progress* (where 'progress' is to be understood as the defender of a certain rule understands it, thus differently for different social and professional groups). If one does not want to cut oneself off from the means to such progress, only one option remains: one admits that *there simply are no* generally valid rules, that no general

---

[100] Supporters of enlightenment, like Lecky, White, and others, often boast about the good services which reason had allegedly provided during the disposal of the witch craze. This optimistic view of the past does not correspond to the facts. There were arguments, as well as scholars, who could prove witchcraft from experience and simple physical or psychological principles. Irrational influences were the ones which, in the end, knocked off this monstrosity. For a fascinating portrayal, with heaps of literature, see H. R. Trevor-Roper, *The European Witch Craze* (New York: Harper Torchbooks, 1968).

methodology which is independent of history, psychology, physics, belief in God, guides our steps uncompromisingly (this is also the basic thought of all dialectical philosophies). Even the most seemingly trivial demands have their limits and must be given up in certain circumstances. Is it not evident, for instance, that a researcher or a research group can cope better with a problem the more they know about the relevant circumstances? The more knowledge, the more prospects for success. Everyone seems to hold this as a self-evident principle. But it is definitely not self-evident. Too much detail confuses our thoughts and takes away our ability to find simple solutions to complex problems. Is it not evident that a *clear and precise* solution must be preferred to an unclear and ambiguous one? Absolutely not. The 'clear' solution disguises the superficiality and ambiguity which accompany every step of thinking and which make it possible in the first place.[101] One accepts such a result as flawless and turns to other things. Clear solutions are like modern canals, they guide research in a certain direction and shut out alternatives. One does not see further than the walls of one's own canal allow. Third, has it not been determined that we are only allowed to use consistent theories, i.e., is it not imperative to free one's theory from contradictions before one looks at its other features more closely? Absolutely not. In the history of the sciences, an inconsistent theory T″ often follows an inconsistent theory T′ *before* one is all too clear about its logical properties, and this succession would be delayed *ad infinitum* by the search for a consistent and logically flawless form. In addition, striving for logical perfection separates the 'superstructure' of science from predictive, experimental practice, and forces the scientist to re-establish the connection in a highly untidy way (see the example in section 6). The devil disappears from the textbooks, but in return he only feels better in practice.[102] And so on. Every methodological rule which one would want to enforce on practice or science has (because of psychological, historical, sociological, etc. laws) undesired consequences. If one acknowledges these consequences, then one is often forced to restrict the rule, or to dispose of it totally.

---

[101] See the short remarks in footnote 61.

[102] The childish objection that from a contradiction every statement follows, as deployed against the use of inconsistent theories, is still being raised. Firstly, this is not true. Statements do not bubble out of themselves. One has to *derive* them, and if one makes this derivation intelligently enough (more intelligently than for instance a simply programmed computer), then no such problem arises. Secondly, this objection, if it is an objection at all, is an objection against *logic, not* against physics. Logicians just have not yet invented a formal system which would allow us adequately to handle the always inconsistent theories of physics. The logic which they offer to us explodes when it is applied to physics, and is therefore useless. The argument that physics, and not logic, has to be improved sounds like the argument of the inventor of a new medication which only heals sick people if they are totally free of bacteria, but otherwise kills them. Except that a physician would think of arguing in this way!

So how does one proceed in a concrete case? How does one solve a theoretical problem? How does one initiate a political change?

One proceeds as a normal person who wants to solve a certain problem, such as a political problem, proceeds. First one informs oneself. Not too much, but also not too little. (What 'much' and 'little' are depends on the situation, as well as on the peculiarities of one's own thinking processes.) Then, one decides whether one can solve the problem alone, or whether one needs the help of others. If the latter is the case, then the choice of whom is made according to character, intelligence, emotional stability, sex, and the like. The choice is different if one wants to make a free coalition, or a tightly organized assault party. At this point, *it may* be useful to study the debate between Marx and Bakunin, or between Lenin and his more 'liberal' opponents in the party, provided that one has enough time, and that the circumstances (the brains of the little group included) demand such a procedure. Then comes the question of action: should one publish or drop bombs? Should one try to persuade (again there are alternatives here: mass meetings, or house calls) or should one intimidate? Decisions on all these questions must be made in the concrete historical situation in which they were posed, and cannot be anticipated, not even in the most vague and general way. Even the existence of professions with strict ethical and theoretical standards (medicine, physics) does not necessarily lead in a certain direction, since why should one follow physics or medicine, or any other sect, just because it is there? But alluding to past 'success' (atomic bomb, etc.), cannot pacify us, because the relationship between theory and technology and invention is anything but clear (again see section 6).

Then come 'moral' problems like the question whether the truth should always be told, or if it is allowable for (the politician, scientists, etc.) to mislead or to lie. Problems like this presuppose that one takes metaphysical monstrosities like 'the truth', or 'justice', and so on seriously – decisions at every turn. Of modern philosophers, Lenin probably understood the situation best: 'History generally, and the history of revolutions in particular, is always richer in content, more varied, more many-sided, more lively and "subtle" than even the best parties [the best groups, professions, individuals] ... can imagine. From this follow two very important practical conclusions: first, that [a group which wants to fulfil its goals] ... must be able to master *all* forms and sides of social activity, without exception; second, [such a group] must be ready to pass from one to another in the quickest and most unexpected manner'.[103] Timorous and insecure people,

---

[103] V. I. Lenin, *'Left Wing' Communism, an Infantile Disorder* (Beijing: Foreign Language Press, 1965), p. 100. This treatise is an excellent antidote to the puritanical Troglodytes of the 'New Left'.

who only act when a Führer, even an abstract leader like methodological rules, leads them by the hand, do not have much prospect of success. They also contaminate the atmosphere with their willingness to give up their liberty. Methodological rules must be adapted to circumstances and continually invented anew. This enlarges our liberty, human dignity and the prospect of success.

8

How to defend society against science

I want to defend society and its inhabitants from all ideologies, science included. All ideologies must be seen in perspective. One must not take them too seriously. One must read them like fairytales which have lots of interesting things to say but which also contain wicked lies, or like ethical prescriptions which may be useful rules of thumb but which are deadly when followed to the letter.

Now – is this not a strange and ridiculous attitude? Science, surely, was always in the forefront of the fight against authoritarianism and superstition. It is to science that we owe our increased intellectual freedom *vis-à-vis* religious beliefs; it is to science that we owe the liberation of mankind from ancient and rigid forms of thought. Today these forms of thought are nothing but bad dreams – and this we learned from science. Science and enlightenment are one and the same thing – even the most radical critics of society believe this. Kropotkin wants to overthrow all traditional institutions and forms of belief, with the exception of science. Ibsen criticizes the most intimate ramifications of nineteenth-century bourgeois ideology, but he leaves science untouched. Lévi-Strauss has made us realize that Western Thought is not the lonely peak of human achievement it was once believed to be, but he excludes science from his relativization of ideologies. Marx and Engels were convinced that science would aid the workers in their quest for mental and social liberation. Are all these people deceived? Are they all mistaken about the role of science? Are they all the victims of a chimaera?

To these questions my answer is a firm *Yes and No*.

Now, let me explain my answer.

My explanation consists of two parts, one more general, one more specific.

The general explanation is simple. Any ideology that breaks the hold a comprehensive system of thought has on the minds of men contributes to the liberation of man. Any ideology that makes man question inherited beliefs is an aid to enlightenment. A truth that reigns without checks and balances is a tyrant who must be overthrown and any falsehood that can aid us in the overthrow of this tyrant is to be welcomed. It follows that seventeenth- and eighteenth-century science indeed *was* an instrument of liberation and enlightenment. It does not follow that science is bound to *remain* such an instrument. There is nothing inherent in science or in any other ideology that makes it *essentially* liberating. Ideologies can deteriorate and become stupid religions. Look at Marxism. And that the science of today is very different from the science of 1650 is evident at the most superficial glance.

For example, consider the role science now plays in education. Scientific 'facts' are taught at a very early age and in the very same manner in which religious 'facts' were taught only a century ago. There is no attempt to waken the critical abilities of the pupil so that he may be able to see things in perspective. At the universities the situation is even worse, for indoctrination is here carried out in a much more systematic manner. Criticism is not entirely absent. Society, for example, and its institutions, are criticized most severely and often most unfairly and this already at the elementary school level. But science is excepted from the criticism. In society at large the judgement of the scientist is received with the same reverence as the judgement of bishops and cardinals was accepted not too long ago. The move towards 'demythologization', for example, is largely motivated by the wish to avoid any clash between Christianity and scientific ideas. If such a clash occurs, then science is certainly right and Christianity wrong. Pursue this investigation further and you will see that science has now become as oppressive as the ideologies it once had to fight. Do not be misled by the fact that today hardly anyone gets killed for joining a scientific heresy. This has nothing to do with science. It has something to do with the general quality of our civilization. Heretics in science are still made to suffer from the *most severe* sanctions this relatively tolerant civilization has to offer.

But – is this description not utterly unfair? Have I not presented the matter in a very distorted light by using tendentious and distorting terminology? Must we not describe the situation in a very different way? I have said that science has become *rigid*, that it has ceased to be an instrument of *change* and *liberation* without adding that it has found the *truth*, or a large part thereof. Considering this additional fact we realize, so the objection goes, that the rigidity of science is not due to human wilfulness. It lies in the nature of things. For once we have discovered the truth – what else can we do but follow it?

This trite reply is anything but original. It is used whenever an ideology wants to reinforce the faith of its followers. 'Truth' is such a nicely neutral word. Nobody would deny that it is commendable to speak the truth and wicked to tell lies. Nobody would deny that – and yet nobody knows what such an attitude amounts to. So it is easy to twist matters and to change allegiance to truth in one's everyday affairs into allegiance to the Truth of an ideology which is nothing but the dogmatic defence of that ideology. And it is of course *not* true that we *have* to follow the truth. Human life is guided by many ideas. Truth is one of them. Freedom and mental independence are others. If Truth, as conceived by some ideologists, conflicts with freedom then we have a *choice*. We may abandon freedom. But we may also abandon Truth. (Alternatively, we may adopt a more sophisticated idea of truth that no longer contradicts freedom; that was Hegel's solution.) My criticism of modern science is that it inhibits freedom of thought. If the reason is that it has found the truth and now follows it then I would say that there are better things than first finding, and then following such a monster.

This finishes the general part of my explanation.

There exists a more specific argument to defend the exceptional position science has in society today. Put in a nutshell the argument says (1) that science has finally found the correct *method* for achieving results and (2) that there are many *results* to prove the excellence of the method. The argument is mistaken – but most attempts to show this lead into a dead end. Methodology has by now become so crowded with empty sophistication that it is extremely difficult to perceive the simple errors at the basis. It is like fighting the hydra – cut off one ugly head, and eight formalizations take its place. In this situation the only answer is superficiality: when sophistication loses content then the only way of keeping in touch with reality is to be crude and superficial. This is what I intend to be.

AGAINST METHOD

There is a method, says part (1) of the argument. What is it? How does it work?

One answer which is no longer as popular as it used to be is that science works by collecting facts and inferring theories from them. The answer is unsatisfactory as theories never *follow* from facts in the strict logical sense. To say that they may yet be *supported* by facts assumes a notion of support that (a) does not show this defect and is (b) sufficiently sophisticated to permit us to say to what extent, say, the theory of relativity is supported by the facts. No such notion exists today nor is it likely that it will ever be found (one of the problems is that we need a notion of support in which grey ravens can be said to support 'All ravens are black'). This was realized by conventionalists and transcendental idealists who pointed out that theories *shape* and *order* facts and can therefore be retained come what may. They can be retained because the human mind either consciously or unconsciously carried out its ordering function. The trouble with these views is that they assume for the mind what they want to explain for the world, viz. that it works in a regular fashion. There is only one view which overcomes all these difficulties. It was invented twice in the nineteenth century, by Mill, in his immortal essay *On Liberty*, and by some Darwinists who extended Darwinism to the battle of ideas. This view takes the bull by the horns: theories cannot be justified and their excellence cannot be shown without reference to other theories. We may explain the *success* of a theory by reference to a more comprehensive theory (we may explain the success of Newton's theory by using the general theory of relativity); and we may explain our *preference* for it by comparing it with other theories. Such a comparison does not establish the intrinsic excellence of the theory we have chosen. As a matter of fact, the theory we have chosen may be pretty lousy. It may contain contradictions, it may conflict with well-known facts, it may be cumbersome, unclear, ad hoc in decisive places and so on. But it may still be better than any other theory that is available at the time. It may in fact be the best lousy theory there is. Nor are the standards of judgement chosen in an absolute manner. Our sophistication increases with every choice we make, and so do our standards. Standards compete just as theories compete and we choose the standards most appropriate to the historical situation in which the choice occurs. The rejected alternatives (theories; standards; 'facts') are not eliminated. They serve as correctives (after all, we may have made the wrong choice) and they also explain the content of the preferred views (we understand relativity better when we understand the structure of its competitors; we know the full meaning of freedom only if we have an idea of life in a totalitarian state, of its advantages – and there are many advantages – as well as of its disadvantages). Knowledge so conceived is an ocean of alternatives channelled and subdivided by an ocean of standards. It forces our mind to make imaginative choices and thus makes it grow. It makes our mind capable of choosing, imagining, criticizing.

Today this view is often connected with the name of Karl Popper. But there are some very decisive differences between Popper and Mill. To start with, Popper developed his view to solve a special problem of epistemology – he wanted to solve 'Hume's problem'. Mill, on the other hand, is interested in conditions favourable to human growth. His epistemology is the result of a certain theory of man, and not the other way around. Also, Popper, being influenced by the Vienna Circle, improves on the logical form of a theory before discussing it while Mill uses every theory in the form in which it occurs in science. Thirdly, Popper's standards of comparison are rigid and fixed while Mill's standards are permitted to change with the historical situation. Finally, Popper's standards eliminate competitors once and for all: theories that are either not falsifiable, or falsifiable and falsified have no place in science. Popper's criteria are clear, unambiguous, precisely formulated; Mill's criteria are not. This would be an advantage if science itself were clear, unambiguous, and precisely formulated. Fortunately, it is not.

To start with, no new and revolutionary scientific theory is ever formulated in a manner that permits us to say under what circumstances we must regard it as endangered: many revolutionary theories are unfalsifiable. Falsifiable versions do exist, but they are hardly ever in agreement with accepted basic statements: every moderately interesting theory is falsified. Moreover, theories have formal flaws, many of them contain contradictions, ad hoc adjustments, and so on and so forth. Applied resolutely, Popperian criteria would eliminate science without replacing it by anything comparable. They are useless as an aid to science.

In the past decade this has been realized by various thinkers, Kuhn and Lakatos among them. Kuhn's ideas are interesting but, alas, they are much too vague to give rise to anything but lots of hot air. If you don't believe me, look at the literature. Never before has the literature on the philosophy of science been invaded by so many creeps and incompetents. Kuhn encourages people who have no idea why a stone falls to the ground to talk with assurance about scientific method. Now I have no objection to incompetence but I do object when incompetence is accompanied by boredom and self-righteousness. And this is exactly what happens. We do not get interesting false ideas, we get boring ideas or words connected with no ideas at all. Secondly, wherever one tries to make Kuhn's ideas more definite one finds out that they are *false*. Was there ever a period of normal science in the history of thought? No – and I challenge anyone to prove the contrary.

Lakatos is immeasurably more sophisticated than Kuhn. Instead of theories he considers research programmes which are sequences of theories connected by methods of modification, so-called heuristics. Each theory in the sequence may be full of faults. It may be beset by anomalies, contradictions, ambiguities. What counts is not the shape of the single theories, but the tendency exhibited by the sequence. We judge historical developments, achievements over a period of time, rather than the situation at a particular time. History and methodology are combined into a single enterprise. A research programme is said to progress if the sequence of theories leads to novel predictions. It is said to degenerate if it is reduced to absorbing facts that have been discovered without its help. A decisive feature of Lakatos' methodology is that such evaluations are no longer tied to methodological rules which tell the scientist either to retain or to abandon a research programme. Scientists may stick to a degenerating programme, they may even succeed in making the programme overtake its rivals and they therefore proceed rationally with whatever they are doing (provided they continue calling degenerating programmes degenerating and progressive programmes progressive). This means that Lakatos offers *words* which *sound* like the elements of a methodology; he does not offer a methodology. There is no method according to the most advanced and sophisticated methodology in existence today. This finishes my reply to part (1) of the specific argument.

According to part (2), science deserves a special position because it has produced *results*. This is an argument only if it can be taken for granted that nothing else has ever produced results. Now it may be admitted that almost everyone who discusses the matter makes such an assumption. It may also be admitted that it is not easy to show that the assumption is false. Forms of life different from science have either disappeared or have degenerated to an extent that makes a fair comparison impossible. Still, the situation is not as hopeless as it was only a decade ago. We have become acquainted with methods of medical diagnosis and therapy which are effective (and perhaps even more effective than the corresponding parts of Western medicine) and which are yet based on an ideology that is radically different from the ideology of Western science. We have learned that there are phenomena such as telepathy and telekinesis which are obliterated by a scientific approach and which could be used to do research in an entirely novel way (earlier thinkers such as Agrippa of Nettesheim, John Dee, and even Bacon were aware of these phenomena). And then – is it not the case that the Church saved souls while science often does the very opposite? Of course, nobody now believes in the ontology that underlies this judgement. Why? Because of ideological pressures identical with those which today make us listen to science to the exclusion of everything else. It is also true that phenomena such as telekinesis and acupuncture may eventually be absorbed into the body of science and may therefore be called 'scientific'. But note that this happens only *after* a long period of resistance during which a science *not yet* containing the phenomena wants to get the upper hand over forms of life that contain them. And this leads to a further objection against part (2) of the specific argument. The fact that science has results counts in its favour only if these results were achieved by science alone, and without any outside help. A look at history shows that science hardly ever gets its results in this way. When Copernicus introduced a new view of the universe, he did not consult *scientific* predecessors, he consulted a crazy Pythagorean such as Philolaos. He adopted his ideas and he maintained them in the face of all sound rules of scientific method. Mechanics and optics owe a lot to artisans, medicine to midwives and witches. And in our own day we have seen how the interference of the state can advance science: when the Chinese communists refused to be intimidated by the judgement of experts and ordered traditional medicine back into universities and hospitals there was an outcry all over the world that science would now be ruined in China. The very opposite occurred: Chinese science advanced and Western science learned from it. Wherever we look we see that great scientific advances are due to outside interference which is made to prevail in the face of the most basic and most 'rational' methodological rules. The lesson is plain: there does not exist a single argument that could be used to support the exceptional role which science today plays in society. Science has done many things, but so have other ideologies. Science often proceeds systematically, but so do other ideologies (just consult the records of the many doctrinal debates that took place in the Church) and, besides, there are no overriding rules which are adhered to under any circumstances; there is no 'scientific methodology' that can be used to separate science from the rest. *Science is just one of the many ideologies that propel society and it should be treated as such* (this statement applies even to the most progressive and most dialectical sections of science). What consequences can we draw from this result?

The most important consequence is that there must be a *formal separation between state and science* just as there is now a formal separation between state and church. Science may influence society but only to the extent to which any political or other pressure group is permitted to influence society. Scientists may be consulted on important projects but the final judgement must be left to the democratically elected consulting bodies. These bodies will consist mainly of laymen. Will the laymen be able to come to a correct judgement? Most certainly, for the competence, the complications and the successes of science are vastly exaggerated. One of the most exhilarating experiences is to see how a lawyer, who is a layman, can find holes in the testimony, the technical testimony of the most advanced expert and thus prepare the jury for its verdict. Science is not a closed book that is understood only after years of training. It is an intellectual discipline that can be examined and criticized by anyone who is interested and that looks difficult and profound only because of a systematic campaign of obfuscation carried out by many scientists (though, I am happy to say, not by all). Organs of the state should never hesitate to reject the judgement of scientists when they have reason for doing so. Such rejection will educate the general public, will make it more confident and it may even lead to improvement. Considering the sizeable chauvinism of the scientific establishment we can say: the more Lysenko affairs the better (it is not the *interference* of the state that is objectionable in the case of Lysenko, but the *totalitarian* interference which kills the opponent rather than just neglecting his advice). Three cheers <for> the fundamentalists in California who succeeded in having a dogmatic formulation of the theory of evolution removed from the text books and an account of Genesis included (but I know that they would become as chauvinistic and totalitarian as scientists are today when given the chance to run society all by themselves. Ideologies are marvellous when used in the company of other ideologies. They become boring and doctrinaire as soon as their merits lead to the removal of their opponents). The most important change, however, will have to occur in the field of *education*.

EDUCATION AND MYTH

The purpose of education, so one would think, is to introduce the young into life, and that means: into the *society* where they are born and into the *physical universe* that surrounds the society. The method of education often consists in the teaching of some *basic myth*. The myth is available in various versions. More advanced versions may be taught by initiation rites which firmly implant them into the mind. Knowing the myth the grown-up can explain almost everything (or else he can turn to experts for more detailed information). He is the master of Nature and of Society. He understands them both and he knows how to interact with them. However, *he is not the master of the myth that guides his understanding.*

Such further mastery was aimed at, and was partly achieved, by the Presocratics. The Presocratics not only tried to understand the *world*. They also tried to understand, and thus to become the masters of, the *means of understanding the world.* Instead of being content with a single myth they developed many and so diminished the power which a well-told story has over the minds of men. The sophists introduced still further methods for reducing the debilitating effect of interesting, coherent, 'empirically adequate' etc. etc. tales. The achievements of these thinkers were not appreciated and they certainly are not understood today. When teaching a myth, we want to increase the chance that it will be understood (i.e. no puzzlement about any feature of the myth), believed, *and accepted*. This does not do any harm when the myth is counterbalanced by other myths: even the most dedicated (i.e. totalitarian) instructor in a certain version of Christianity cannot prevent his pupils from getting in touch with Buddhists, Jews and other disreputable people. It is very different in the case of science, or of rationalism where the field is almost completely dominated by the believers. In this case it is of paramount importance to strengthen the minds of the young and 'strengthening the minds of the young' means strengthening them *against* any easy acceptance of comprehensive views. What we need here is an education that makes people *contrary, counter-suggestive without* making them incapable of devoting themselves to the elaboration of any single view. How can this aim be achieved?

It can be achieved by protecting the tremendous imagination which children possess and by developing to the full the spirit of contradiction that exists in them. On the whole children are much more intelligent than their teachers. They succumb, and give up their intelligence because they are bullied, or because their teachers get the better of them by emotional means. Children can learn, understand and keep separate two to three different languages ('children' and by this I mean three to five-year-olds, *not* eight-year-olds who were experimented upon quite recently and did not come out too well; why? because they were already loused up by incompetent teaching at an earlier age). Of course, the languages must be introduced in a more interesting way than is usually done. There are marvellous writers in all languages who have told marvellous stories – let us begin our language teaching with *them* and not with 'der Hund hat einen Schwanz' and similar inanities. Using stories we may of course also introduce 'scientific' accounts, say, of the origin of the world and thus make the children acquainted with science as well. But science must not be given any special position except for pointing out that there are lots of people who believe in it. Later on the stories which have been told will be supplemented with 'reasons' where by reasons I mean further accounts of the kind found in the tradition to which the story belongs. And, of course, there will also be contrary reasons. Both reasons and contrary reasons will be told by the experts in the fields and so the young generation becomes acquainted with all kinds of sermons and all types of wayfarers. It becomes acquainted with them, it becomes acquainted with their stories and every individual can make up his mind which way to go. By now everyone knows that you can earn a lot of money and respect and perhaps even a Nobel Prize by becoming a scientist, so, many will become scientists. They will *become* scientists *without having been taking in by the ideology of science*, they will *be* scientists *because they have made a free choice*. But has not much time been wasted on unscientific subjects and will this not detract from their competence once they have become scientists? Not at all! The progress of science, of good science, depends on novel ideas and on intellectual freedom: science has very often been advanced by outsiders (remember that Bohr and Einstein regarded themselves as outsiders). Will not many people make the wrong choice and end up in a dead end? Well, that depends on what you mean by a 'dead end'. Most scientists today are devoid of ideas, full of fear, intent on producing some paltry result so that they can add to the flood of inane papers that now constitutes 'scientific progress' in many areas. And, besides, what is more important? To lead a life which one has chosen with open eyes, or to spend one's time in the nervous attempt of avoiding what some not so intelligent people call 'dead ends'? Will not the number of scientists decrease so that in the end there is nobody to run our precious laboratories? I do not think so. Given a choice many people may choose science, for a science that is run by free agents looks much more attractive than the science of today which is run by slaves, slaves of institutions and slaves of 'reason'. And if there is a temporary shortage of scientists the situation may always be remedied by various kinds of incentives. Of course, scientists will not play any predominant role in the society I envisage. They will be more than balanced by magicians, or priests, or astrologers. Such a situation is unbearable for many people, old and young, right and left. Almost all of you have the firm belief that at least *some* kind of truth has been found, that it must be preserved, and that the method of teaching I advocate and the form of society I defend will dilute it and make it finally disappear. You have this firm belief; many of you may even have reasons. *But what you have to consider is that the absence of good contrary reasons is due to a historical accident;* it does *not* lie in the nature of things. Build up the kind of society I recommend and the views you now despise (without knowing them, to be sure) will return in such splendour that you will have to work hard to maintain your own position and will perhaps be entirely unable to do so. You do not believe me? Then look at history. Scientific astronomy was firmly founded on Ptolemy and Aristotle, two of the greatest minds in the history of Western Thought. Who upset their well argued, empirically adequate and precisely formulated system? Philolaos the mad and antediluvian Pythagorean. How was it that Philolaos could stage such a comeback; because he found an able defender: Copernicus. Of course, you may follow your intuitions as I am following mine. But remember that your intuitions are the result of your 'scientific' training where by science I also mean the science of Karl Marx. My training, or, rather, my non-training, is that of a journalist who is interested in strange and bizarre events. Finally, is it not utterly irresponsible, in the present world situation, with millions of people starving, others enslaved, down-trodden, in abject misery of body and mind, to think luxurious thoughts such as these? Is not freedom of choice a luxury under such circumstances: is not the flippancy and the humour I want to see combined with the freedom of choice a luxury under such circumstances? Must we not give up all self-indulgence and *act*? Join together, and *act*? That is the most important objection which today is raised against an approach such as the one recommended by me. It has tremendous appeal, it has the appeal of unselfish dedication. Unselfish dedication – to what? Let us see!

We are supposed to give up our selfish inclinations and dedicate ourselves to the liberation of the oppressed. And selfish inclinations are what? They are our wish for maximum liberty of thought in the society in which we live *now*, maximum liberty not only of an abstract kind, but expressed in appropriate institutions and methods of teaching. This wish for concrete intellectual and physical liberty in our own surroundings is to be put aside, for the time being. This assumes, first, that we do not need this liberty for our task. It assumes that we can carry out our task with a mind that is firmly closed to some alternatives. It assumes that the correct way of liberating others *has already been found* and that all that is needed is to carry it out. I am sorry, I cannot accept such doctrinaire self-assurance in such extremely important matters. Does this mean that we cannot act at all? It does not, but it means that *while acting we have to try to realize as much of the freedom I have recommended so that our actions may be corrected in the light of the ideas we get while increasing our freedom.* This will slow us down, no doubt, but are we supposed to charge ahead simply because some people tell us that they have found an explanation for all the misery and an excellent way out of it? Also we want to liberate people not to make them succumb to a new kind of slavery, *but to make them realize their own wishes,* however different these wishes may be from our own. Self-righteous and narrow-minded liberators cannot do this. As a rule they soon impose a slavery that is worse, because more systematic, than the very sloppy slavery they have removed. Why would anyone want to liberate anyone else? Surely not because of some *abstract* advantage of liberty but because liberty is the best way to free development *and thus to happiness.* We want to liberate people *so that they can smile.* Shall we be able to do this if we ourselves have forgotten how to smile and are frowning on those who still remember? Shall we then not spread another disease, comparable to the one we want to remove, the disease of puritanical self-righteousness? Do not object that dedication and humour do not go together – Socrates is an excellent example to the contrary. *The hardest task needs the lightest hand or else its completion will not lead to freedom but to a tyranny much worse than the one it replaces.*

9
Let's make more movies

In the first scene of Brecht's *Life of Galileo,* Galileo uses a short demonstration to convince the boy Andrea of the relativity of motion. In scene 7 he repeats the point for a learned cardinal. In scene 9 he refutes some of Aristotle's views on floating bodies by a simple and elegant experiment.

When realized on the stage these brief episodes make us acquainted with some features of a scientific debate. A few more examples, and we might know how to argue in similar cases. But they also show how people *behave* when engaged in argument; how their behaviour influences the life of others, and what role such influence plays in society. Presented swiftly, concisely, and forcefully, the episodes impose upon us an interesting and uncomfortable conflict: having been trained by our teachers, by the pressure of professions, and by the general climate of a liberal-scientific age to 'listen to reason', we quite automatically abstract from 'external circumstances' and concentrate on the *logic* of a demonstration. A good play, on the other hand, does not permit us to overlook faces, gestures – or what one might call the *physiognomy* of an argument. A good play uses the *physical manifestations of reason* to irritate our senses and disturb our feelings so that they get in the way of a smooth and 'objective' appraisal. It tempts us to judge an event by the interplay of *all* the agencies that cause its occurrence. Even better, a good play does not merely tempt us; it deflects us from our intention to use rational criteria only; it gives the material manifestations of the idea business a chance of making an impression, and it thus forces us to *judge reason* rather than use it as a basis for judging everything else. Let us see how this works in a special case.

Brecht's Galileo is not a professional. The fact that he has ideas and can support them by argument is the least important thing about him. What interests the writer is that Galileo is *a new type of thinker,* that he is a man rather than a 'trained scientist' (pp. 48, 106).[1] He is virile, sensual, impetuous, aggressive, extremely curious, almost a voyeur, a glutton physically and intellectually (p. 63), and a born showman (p. 41). When the curtain rises we see him half-naked, enjoying morning bath, breakfast, astronomical conversation – all at the same time. Thinking is for him a joyful and libidinous activity, the play of his hands in his pockets that accompanies it and expresses its emotional nature 'approaches the limits of the obscene' (p. 51). This is the man who explains Copernicus to Andrea 'in an offhand manner' and without trying to drive the point home. He simply 'leaves [the boy] alone with his thoughts' (p. 51). He leaves him alone not because of lack of interest, for Andrea, despite his youth *and despite his ignorance* is treated as an equal ('as a result of our research, Signora Sarti, Andrea and I have found, after long debate ...' (p. 1236)). Nor is the collaboration enforced; it is the natural result of a charming friendship between a vigorous scholar and an intelligent, inquisitive, and headstrong boy. Thought, so it seems, has left university and monastery and has become part of everyday life. This is the situation Brecht wants to discuss.

The situation is not unambiguous. We are not merely shown a new form of life, we are also shown some of its internal contradictions and the problems to which they lead.

For example, Galileo is fond of certain phrases, gestures; he uses them frequently and occasionally with an air of self-righteousness. Andrea repeats them, though less imaginatively and much more rigidly. When the situation seems to get out of control, when the discoveries of his master are in danger of being pushed aside, then all he can do is describe them with raised voice (p. 1327). In the end he turns out to be a somewhat unintelligent and slightly unstable Puritan (Scene 14). Could it be that relaxed collaboration creates slaves more readily than does the usual teacher-pupil interaction with its emphasis on training and domination? Galileo's daughter who wants to participate in what seems to be such an entertaining life is cruelly rejected – 'this is not a toy...' (p. 1258) – so the new knowledge business that announces itself with the nude Galileo in Scene 1 is not accessible to everyone, nor is it free from stereotype. The distinction between those who play in the correct fashion and those who do not is driven home when Galileo confronts Mucius, who has gone his own way. He sees him 'with his pupils crowded behind him' (p. 1299) like a pack of unsure dogs. The dogs not only protect their master; they also want to be fed and amused; and Galileo, who does not always come up to their

---

[1] In what follows I am quoting or paraphrasing from *Materialen zu Brechts 'Leben des Galilei'* (Frankfurt: Suhrkamp Verlag, 1967), pp. 1–212, and *Gesammelte Werke 3* (Frankfurt: Suhrkamp Verlag, 1967), pp. 1230 ff.

rather narrow moral expectations, resorts to tricks to keep them interested and loyal, to 'suppress their discontent' (p. 63). The tricks he produces are important scientific demonstrations, they are essential parts of what we call, in retrospect, the 'scientific revolution', they are full of deep insight, and they are performed with an elegance and ease that make them veritable works of art (p. 62). Yet their origin now almost seems to be the wish to dominate, not by physical power, not by fear, but by the much more subtle and vicious power of truth. And their function: to satisfy the intellectual greed of his followers and to tie them closer to him. (Politicians need new wars, and scientists new discoveries to prevent their soldiers from becoming discontented.)

It is quite true. Research *has* ceased to be a purely contemplative process; it *has* become part of the physical world; it *has* started to influence people in new ways; it *has* established new relations between them. But instead of becoming an instrument of liberation as well, it creates new needs which are as insatiable as the needs of a sexual pervert: '[Galileo] refers to his unsatisfied drive to do research in the very same manner in which an arrested sex maniac might refer to his glands' (p. 60). Even the happiness of his daughter, her whole life, counts little when it conflicts with the urge to know (p. 1312).

In the play this aspect of the new science is explained by Galileo's political failure. Research goes on afterward. The results are more splendid than ever. They are still revolutionary, from the point of view of mechanics and astronomy, but they have lost their chance to reform society for a long time to come. Knowledge is a secret for professionals again; the content has changed; the form remains. This is what the story *tells* us. In addition it *shows* that this particular aspect was present from the beginning and thus exhibits the contradictory nature of every historical event.

So far, a brief and very incomplete sketch of the working of a tiny part of a complex and colourful machinery. What can we learn from it?

III

The problem that appears in the play is one of the most important *philosophical problems*. It is the problem of the role of reason in society and in our private lives, and of the changes which reason undergoes in the course of history. What happens when a strange and ethereal entity such as thought that has 'eternal' laws of its own and makes submission to these laws a condition of rationality, knowledge, progress, even of humanity, takes up residence in the physical universe and starts directing the lives of men? Are the consequences always desirable, and what changes should be carried out if they are not? On stage the problem is not dealt with in a purely conceptual manner. It is *shown* as much as it is *explained*. This is anything but a disadvantage. Philosophical discussion has often been criticized for being too abstract, and one has demanded that the analysis of concepts such as *reason, thought, knowledge*, etc., be tied to concrete examples. Now concrete examples are circumstances which guide the application of a term and give content to the corresponding concept. The theatre not only provides such circumstances, it also arranges them in a way that *inhibits* the facile progression of abstractions and forces us to reconsider the most familiar conceptual connections. Also the business of speculation which occasionally seems to swallow everything else is here set off from a rich and changing visual background that reveals its limitations and helps us to judge it as a whole. (Today an interesting visual point arises from the fact that businessmen, philosophers, scientists, and hired killers all dress alike and have comparable professional standards. But the briefcase in the hands of these pillars of democracy may contain a contract, a thesis, a new calculation of the S-matrix, or a submachine gun.) It is of course possible to present the additional elements in words, but only at the expense of regarding our problem as solved before we have started examining it. For we now simply *assume* that everything can be translated into the medium of ideas. We have to conclude, then, that there are better ways of dealing with philosophical problems than verbal exchange, written discourse, and, *a fortiori*, scholarly research.

IV

This result was well known at a time when philosophy was still close enough to the arts and to myth to be able to avoid the trap of intellectualism. Plato's objections to writing (*Phaedrus* 275a ff.); his use of dialogue as a means of bringing in apparently extraneous material; his frequent changes of style (*Philebus* 23b); his refusal to develop a precise and standardized language, a jargon (*Theaetetus* 184c); and, above all, his appeal to myth in places where a modern philosopher would expect a scintillating culmination of argumentative skill – all these features show that he was aware of the limitations of a purely conceptual approach. Earlier societies (and some non-industrial cultures of today) have overcome these limitations in a different way, not by trying to *rebuild* emotions, gestures, physical phenomena in the medium of language, but by *making them part* of the basic ideology. This ideology represents the entire cosmos, and it uses all the resources of society – architecture, thought, dance, music, dreams, drama, medicine, education, even the most pedestrian activity in the process.[2] Philosophy, however, chose to restrict itself to the word.

---

[2] <See> M. Griaule, *Conversations with Ogotemmeli* (Oxford: Oxford University Press, 1965).

V

This restriction was soon followed by others. Plato's attempt to create an art form that could be used to *talk* about reason *and* to *show* its clash with the 'world of appearances' was not continued. Technical terminology, standardized arguments replaced his colourful and imprecise language; the treatise replaced the dialogue; the development of ideas became the only topic. For a while one tried to construct comprehensive conceptual systems and used them for evaluating the relative merits of institutions, professions, results. There was a hierarchy of professions; each individual subject received meaning from the total structure and provided a content for it. The hierarchy fell apart with the demand for autonomy that arose in the fifteenth and sixteenth centuries and became orthodoxy with the arrival of modern science. Even philosophy was broken up into various disciplines with special problems that had little relation to each other. Was its quality improved? It was not, as is shown by the history of one of its more desiccated parts, viz., the *philosophy of science.*

VI

The scientific revolution of the sixteenth and seventeenth centuries does not yet suffer from the effects of specialization. Science and philosophy are still closely related. Philosophy is used to expose and to remove the hardened dogmas of the schools and it plays a most important role in the arguments about the Copernican system, in the development of optics, and in the construction of a non-Aristotelian dynamics. Almost every work of Galileo – the real Galileo and not Brecht's invention – is a mixture of philosophical, mathematical, physical, psychological ideas which collaborate without giving the impression of incoherence. This is the *heroic time* of the philosophy of science. It is not content just to *mirror* a science that develops independently of it; nor is it so distant as to deal with alternative *philosophies* only. It *builds* science, and defends it against resistance and explains its consequences.

Now it is interesting to see how this active and critical enterprise is gradually replaced by a more conservative creed that has technical problems of its own, and how there arises a new subject that accompanies science and comments on it but it refrains from interfering. The development is occasionally interrupted by a vigorous and irrepressible thinker such as Ernst Mach, who sets his ideas against the well-established mechanical world view of the nineteenth century and who wants to change science not just to increase its efficiency, but also to preserve freedom of thought.

His suggestions are taken up by scientists and philosophers. The former use them in the Galilean manner, to awaken science from its dogmatic slumber and to turn it upside down. The result in *philosophy* is a new conformism. In the beginning, this conformism has all the appearances of a Great Revolution: 'metaphysical' philosophies are criticized, sneered at, or simply pushed aside; weak speculation in the sciences is triumphantly exposed (not without considerable help from the scientists themselves); advances in logic are turned into formidable machines of war. But now, after all this initial commotion has subsided, what remains?

There remains a subject whose professed aim is to 'explicate' science, which means we are not supposed to change science, but to make it clearer. The call for clarity is raised without any attention to the problems of the scientist. Satisfaction of the demands of a particular school-philosophy, namely, logical empiricism, is deemed sufficient. What we have here is therefore a *double conformism: both* science *and* logical empiricism are to be preserved, and 'explication' is the machinery that does the dirty work. Only this machinery soon gets entangled with itself (paradox of confirmation, counterfactuals, grue), so that the main problem is now its own survival and not the embalming of science and of positivism. That this struggle for survival is interesting to watch I am the last one to deny. What I do deny is that physics or biology or psychology or even philosophy can profit from participating in it. It is much more likely that they will be *retarded*.

They will be retarded because of the naïve simplicity of the philosophers' approach and because of its mistaken urge for precision. After all, we are not only interested in whether a given methodology solves problems that appear when certain simple logical models are used, or whether it agrees with the principles of a popular ideology such as logical empiricism. We also want to know whether it has a point of attack in the knowledge we possess – *and that means* in the imperfect, internally inconsistent, unfinished, vague, incoherent, ambiguous theories 'facts' we happen to accept at a certain time – and how we can improve *this* knowledge in the complex physical, psychological, social conditions in which science finds itself. A logically perfect set of rules may have disastrous consequences when applied in practice (a logically perfect idea of dancing may cause recurrent cramps); or, what is more likely, it may turn out to be absolutely useless.

Such a judgement can of course be obtained only by putting philosophy in a wider context and by combining methodological speculation with historical inquiry. This was done not so long ago, and the results are amazing: science violates *all* the conditions which logical empiricists pretend to have abstracted from it, and the attempt to enforce the conditions would wipe it out without putting anything comparable in its place.[3] The

---

[3] <See> my 'Against Method: Outline of an Anarchistic Theory of Knowledge' in M. Radner and S. Winokur (eds.), *Minnesota Studies in the Philosophy of Science, volume IV* (Minneapolis: University of Minnesota Press, 1970).

separation of science and the philosophy of science has indeed become complete. What is the remedy?

VII

In the case of science versus the philosophy of science the remedy is obvious. What is needed is a philosophy that does not just comment from the outside, but participates in the process of science itself. There must not be any boundary line between science and philosophy. Nor should one be content with an increase in efficiency, truth content, empirical content, or what have you. All these things count little when compared with a happy and well-rounded life. We need a philosophy that gives man the power and the motivation to make science more civilized rather than permitting a superefficient, supertrue, but otherwise barbaric science to debase man. Such a philosophy must show and examine all the consequences of a particular form of life including those which cannot be presented in words. Thus there must not be any boundary between philosophy and the rest of human life either.

We must rid ourselves of the restriction to words, treatises, and scholarship that has shaped philosophy for now well over two thousand years. We must try to revive mythical ways of presentation and we must also try to adapt them to contemporary needs and resources. This brings us back to the problems at the beginning of the essay.

VIII

One of the characteristics of ancient myth is that the elements which it uses to represent the cosmos and the role of man in it are arranged to increase the stability of the whole. Each part is related to each other part in a way that guarantees the eternal survival of the society, and of the state of mind it represents. This is not always an advantage. We want to improve the quality of life and we want to be able to see where improvement is needed. Now discontent arises only where parts are in conflict with each other, for example, when one's wishes and emotions are found to conflict with external reality. New *ideas* arise when the possibility of such a conflict is not excluded. A comprehensive system of presentation is potentially progressive only when its parts can be set against each other. And parts can be set against each other only when they have first been *separated* from the whole and permitted to live their own lives. The separation of subjects that is such a pronounced characteristic of modern philosophy is therefore not altogether undesirable. It is a step on the way to a more satisfactory type of myth. What is needed to proceed further is not the return to harmony and stability as too many critics of the *status quo*, Marxists included, seem to think, but a form of life in which the constituents of older myths – theories, books, images, emotions, sounds, institutions – enter as interacting *but antagonistic* elements. Brecht's theatre was an attempt to create such a form of life. He did not entirely succeed. I suggest we try movies instead.

IX

One of the advantages of film is that the number of the elements which are at the disposal of the director and their degrees of freedom are vastly greater than in any other medium. On the stage it is impossible to separate colour and object and to show their effects independently. The film can overcome this difficulty. On the stage it is impossible to separate expression from the presence of a human body. The film can overcome this impossibility. On the stage it is impossible to show how a character is put together piece by piece until a strong and vigorous individual stands before us. The film can overcome this impossibility. Of course, the impossibilities of the theatre I have just described are a matter of degree, and they are not absolute. I shall never forget how Ekkehard Schall,[4] step by step, transformed the character of Arturo Ui. Each step was a superb exercise in slapstick, the intermediate results were utterly laughable until out of their mere accumulation there suddenly emerged a hideous shape of incredible political force. The theatre is much richer than the average critic is inclined to think. But the film still adds to it without losing its achievements. It can show the transformation of *faces* (operation; makeup; mimicking) in addition to the transformation of bodies. It can show the effects of distance in space, time, and context. It can move from stage or book into life and back again into stage and book. And so on. Of course, it will need a new generation of *thinking* directors to exploit all the possibilities of this medium. But their rise will be the beginning of mythologies that will continue the work of the older philosophers and put an end to the strange business that has lived off their results in the last few centuries.

[4] [Brecht's son-in-law (ed.).]

10

Rationalism, relativism and scientific method

(1) That it is excellent to be rational is admitted by many people, but hardly anyone is able to tell us what it means to be rational and why being rational is so important. The Presocratics were called rational because they omitted the gods from their explanations, the Church Fathers were called rational because they eliminated Gnosticism, Einstein was called rational because he abolished, or seemed to abolish, the aether. In all these cases there is the assumption that some doctrines are true, others false, and being rational means accepting what is believed to be true.

(2) But the truth of a doctrine is not easy to ascertain. The assumption that a certain view is true may turn out to be mistaken. One may even find that the view does not make sense. This applies not only to complex views such as Newton's views of space, time and matter, it applies also to such simple and apparently fundamental principles as the principle of contradiction (contested by Hegelians) and the principle of the excluded middle (contested by constructivists). The realization that all our knowledge has this precarious, 'hypothetical' character makes it rational to explore views assumed to be false. Hence, rationality can no longer be defined as adherence to a certain view.

The second disadvantage of the view that being rational means accepting what is believed to be true is that it takes the idea of truth for granted. But this idea is a relatively recent product. It arose with the Presocratics, it is absent from Homer.[1]

In Homer we have something very close to the notion of fitting which depends on special circumstances. Homer, accordingly, has not one notion of knowledge, but many, without there being any possibility of regarding them as instances of a more comprehensive idea (except by enumeration). Yet there was never any argument to show that the modern and more totalitarian notion of truth has advantages, and what the advantages are.

---

[1] For some comments on this matter and further literature <see> chapter 17 of my essay *Against Method* (London: New Left Books, 1975) (or the improved German version *Wider den Methodenzwang* (Frankfurt: Suhrkamp, 1976)).

(3) It is therefore advisable to connect rationality with procedures rather than with views. And indeed, such a *formal* notion of rationality has become prominent in more recent discussions. Rationality now means acceptance of certain *procedures* (rules, standards) together with the *results* of these procedures, rules, standards; it does not mean acceptance of views (except insofar as the views emerge from the application of the procedures, rules, standards): it is rational (a) to make one's actions conform to certain rules (standards, procedures), i.e. one must not act erratically, and (b) to stick to the procedures, rules, standards that have been chosen.

This explanation of rationality at once raises a whole battalion of questions, for example: why is it better to behave in an orderly fashion rather than erratically? How are the rules that determine rational behaviour to be chosen? How will one determine whether the chosen rules continue to be acceptable and need not be replaced by other rules? And so on.

(4) One answer to the first question that has some chance of being relevant is that the cosmos is an orderly structure which can only be explored by orderly procedures. Neither the assumption nor the consequence can be accepted without criticism. The assumption: that the cosmos is orderly. There are certainly lots of erratic events in it including erratic behaviour on the part of individuals and erratic historical occurrences.[2] Science tries to understand and tame such events on the basis of general principles (sociological laws, etc.). There is no attempt to school the intuition of individuals so that it produces erratic behaviour which is in phase with unusual events. One moves in the opposite direction, behaviour is made more uniform and less capable of dealing with surprises.[3] (This may be one of the reasons why politicians with a theoretical bent are doing such an execrable job. It may also explain why the ability to influence nature in a direct manner has now declined and why parapsychological and paraphysical effects are so difficult to find.)

The consequences of the assumption cannot be accepted either. An orderly world whose laws are not manifest has many surprises in store. None of the apparent regularities it contains is a suitable guide to the laws themselves. And there is no reason to believe that the laws of the world are manifest.

(5) We may classify the attempts to reply to the second and the third question (section 3, last paragraph) by distinguishing naïve and sophisticated rationalism on the one side, cosmological, institutional and normative rationalism on the other:

---

[2] 'Primitive' societies detect and school the ability of children to divine unusual events.
[3] <See> Jean Houston in E. D. Mitchell (ed.) *Psychic Exploration* (New York, 1974), pp. 582ff.

rationalism	cosmological	institutional	normative
naïve			
sophisticated			

The first distinction deals with the *form* of the rules, prescriptions, standards one wants to introduce.

Naïve rationalists assume that there are standards and/or rules which must be obeyed, come what may and which in practice are obeyed by science at its best.

Sophisticated rationalists assume that rules and standards are restricted to certain conditions and that no standards can be presumed to have universal validity. Even the rules of logic may have to be changed when we move from one domain to another. Scientists must keep this in mind and look out for the boundaries.

Naïve rationalism is the philosophy of the founders of Western culture and it comes to the fore in times of crisis and change. Examples are Aristotle, Descartes (but not Bacon), Newton, and Kant. Russell, Popper and Lakatos are more recent examples. Among its ancestors we have the apodictic laws of Exodus and the list of curses embedded in Deuteronomy 27.

Sophisticated rationalism is quite rare. It may be found in traces in Aristotle,[4] it is accepted among some sceptics and it occurs again in Hegel and in dialectical materialism. According to dialectical materialism all principles (standards, rules) have their limits, the contradictions inherent in things drive them towards these limits, and the researcher must not fall behind. Sophisticated rationalists occasionally express their ideas in conditional statements while others have pointed out that making such statements measures of rationality overlooks that they, too, have their limits, that these limits (like all other limits) are discovered by research which therefore cannot be guided by the statements exclusively: concrete research both determines and is determined by standards of rationality. Important ancestors of sophisticated rationalism of the non-dialectical kind are the case laws in the Book of the Covenant (Exodus 21–3) which go back to the Sumerian jurisprudence of the third millennium BC.[5]

(6) The second distinction deals with the *reasons* for the rules, standards, procedures that are being proposed.

Cosmological rationalists view the process of knowledge-building on analogy

---

[4] <See> W. Wieland, *Die Aristotelische Physik* (Tübingen, 1964).
[5] For the distinction between apodictic laws and case laws and their historical ancestors <see> W. F. Albright, *Yahweh and the Gods of Canaan* (New York, 1968), chs. ii and iv.

with physical processes such as the process of bridge-building. The rules of bridge-building involve practical considerations (material, maximum weight, funds), aesthetic considerations (shape of bridge) and facts of nature (including laws and special natural conditions in which the laws are being applied). In the same way the rules of knowledge-building involve practical considerations (funds, wishes of special interest groups, capacity of the computers used, etc.), aesthetic-metaphysical considerations, and facts of nature. Both kinds of rules can be criticized by showing that, given the facts and the aim (to construct a bridge of a certain kind; to improve theories of a certain kind) an application of the rules is not going to lead to the aim. Thus, given a world whose laws are embedded in sizeable fluctuations (which may or may not be reducible to the laws), a principle of falsification that eliminates views inconsistent with facts would lead to a breakdown of knowledge.

(7) *Institutional rationalists* have noticed that the activity of knowledge-building depends on institutions and traditions. So, of course, does the activity of bridge-building. But while the inadequacies of traditions of bridge-building can be ascertained with comparative ease, the inadequacies of epistemic traditions are much harder to find. The reason is not that they are so well hidden, the reason is that epistemic traditions are more pervasive than traditions of bridge-building and therefore allow for a greater variety of adaptations. Thus, in the case of the example of section 6 we may conclude, as we did in that section, that falsification is too severe and that knowledge-construction requires more lenient rules. But we may also conclude that the rule is a suitable guide to natural knowledge and regard the exceptions as miracles.[6] Or we may retain the rule and conclude with the sceptic that knowledge is impossible. Also what counts as a fact, or as a law, depends on criteria of precisely the same kind the cosmological rationalist examines with the help of facts. Considerations such as these have prompted some thinkers to conclude that rules and standards are entirely institutional: we accept them because we participate in certain institutions and traditions and we defend them by reference to these institutions and traditions. The problem is that there are many institutions and traditions – so why choose a particular one as a 'basis of rationality'?

(8) Philosophers of the Enlightenment up to and including Kant have answered this question by asserting that all traditions have certain features in common and that these common features are sufficiently rich and detailed to serve as a basis for rationality:

---

[6] The nova of 1572 and the comet of 1577 were regarded by some as divine interventions and not as refuting instances of the assumption of the unchangeability of the heavens. For the nova, <see> Tycho Brahe *Astronomiae Instauratae Progymnasmata*, for the comet, <see> Doris Hellman, *The Comet of 1577* (New York, 1944), pp. 132, 152, 172.

Und unterm braunen Sud fuhit auch der Hottentot
Die allgemeine Dflicht und der Natur Gebot.[7]

Rules, ideologies, traditions incompatible with such 'laws of nature' are not impossible, they can be discussed, the most beautiful arguments can be used in their favour. However, they are rarely *rich* enough to provide a framework for life in the full sense of the word and they often lack the *strength* to influence such life. We may therefore distinguish between two kinds of traditions and institutions which I shall call primary and secondary traditions (institutions) respectively.[8] Primary traditions contain the ingredients that are necessary for (temporary) survival and understanding. For example, they contain all those principles which make it possible for us to perceive and to understand what we perceive. They arise in a manner that is only in part influenced by reason and that is often difficult to explain: we know only little about perception, we are not even clear about its phenomenological features, let alone its causal ingredients. Secondary traditions have often been built up with the explicit intention of changing (parts of) primary traditions. They are much more intellectual, they rest on principles of reason and are defended by arguments conforming to explicitly formulated rules. It is asserted by the view we are considering at the moment that secondary traditions can never approach the complexity of primary traditions and that they can only change those parts of primary traditions that provide suitable points of attack and levers. Rationality and the conditions for its change are restricted to primary traditions entirely. And there is only one primary tradition.[9] (Hegelians assume that this tradition develops and that all traditions that occur in history and seem to be primary traditions have a place somewhere in this development.)

This theory solves the problem of the institutional rationalist only if its basic assumptions have greater weight than the many traditions that today compete for our attention. This is not the case. Despite determined attempts to find one 'basic rationality' that underlies all societies we are still left with a multiplicity of traditions and institutions of comparable strength and plausibility. Institutional rationalism, therefore, does not solve the problem of rationality.

(9) Only a few people are aware of this situation. Most modern rationalists take their cue either from science or from some 'logical reconstruction of science'. We can easily disregard the latter. Reconstruc-

---

[7] 'And under his brown skin even the Hottentot has a feeling for universal duty and the commandments of nature'. <See> A. O. Lovejoy, 'The Parallel of Deism and Classicism', reprinted in *Essays in the History of Ideas* (Baltimore: Johns Hopkins University Press, 1948), pp. 78ff. The quotation (from Albrecht von Haller's *Über den Ursprung des Ubels*) is on page 87.

[8] 'Tradition' is here used in a wide sense, covering social, psychological as well as bio-physiological phenomena.

[9] This was J. L. Austin's view (private communication). It was also one of the points at issue in the debate between Kant and Herder.

tions arose when philosophers who were unable to participate in scientific debate and unwilling to do without the halo of science turned illiteracy into expertise by insinuating that the simple-minded logical systems they knew revealed deep-lying structural properties of the scientific enterprise. Now there is a simple test to show whether this is indeed the case: replace the part of science that has been 'reconstructed' by the reconstruction and see what happens. In all the cases where the replacement can be carried out the result is clear: science replaced by the reconstruction ceases to work:[10] interesting problems disappear, revolutionary suggestions become either trivial or cease to make sense, concepts lose the ambiguity and indefiniteness that is needed to move from one stage to another. Reconstructions, therefore, cannot replace science as a measure of rationality.

Nor can science itself be such a measure. First, because it lacks the uniformity that is needed to give us a coherent point of view.[11] Secondly, science has frequently employed procedures which are now regarded as 'irrational'. To use it as a standard of rationality we would already have to know how to separate the good from the bad.[12] Thirdly, science is not the only institution that has results, reaches its aims, has a certain amount of coherence. Today, of course, science is believed to be far ahead of alternatives, but this is due to ignorance (of alternatives), arbitrary social decisions, and not to any inherent excellence of science. When modern science arose it had some successes[13] and was close to the heart of powerful interest groups.[14] The successes combined with the power gradually eliminated competitors such as alchemy and the magic world view although these competitors had suffered only a temporary setback and although they were still studied by outstanding scientists such as Newton.[14a] This is a familiar phenomenon: ideas such as the idea of the atomic constitution of

[10]   On the history and the problem of reconstructions <see> my paper 'Against Method: Outline of An Anarchistic Theory of Knowledge' in M. Radner and S. Winokur (eds.), *Minnesota Studies in the Philosophy of Science, volume IV* (Minneapolis: University of Minnesota Press, 1970), my essay 'Die Wissenschaftstheorie – eine bisher unbekannte Form des Irrsinns?' in Kurt Hubner and Albert Menne (eds.), *Natur und Geschichte* (Hamburg: Felix Meiner, 1973), as well as my review of Stegmüller's *The Structure and Dynamics of Theories*, 'Changing Patterns of Reconstruction', *British Journal for the Philosophy of Science*, vol. 28, 1977, pp. 351–369.

[11]   For details <see> *Against Method*, pp. 202ff.

[12]   According to Lakatos we follow the judgement of outstanding scientists. But outstanding scientists have often gone astray and, besides, there is no unambiguous way to separate them from the rest.

[13]   These 'successes' are often a quite mysterious matter. What seems like a success from afar often turns into neutral research, or propaganda when looked at from close by.

[14]   Examples of such groups are described in R. K. Merton, *Science, Technology and Society in Seventeenth Century England* (New York: Howard Fertig, Inc. and Harper and Row, 1970).

[14a] <See> Frank E. Manuel, *The Religion of Isaac Newton* (Oxford: Clarendon Press, 1974), and Betty Jo Teeter Dobbs, *The Foundation of Newton's Alchemy* (Cambridge: Cambridge University Press, 1975).

matter, the idea of the motion of the earth, the idea of action at a distance have their ups and downs, they are occasionally ahead of other ideas, then new evidence turns up, it is not easy to explain this new evidence, so the rivals gain an advantage until the correct explanations are found and the rivals are again overtaken. The particular revolution of the sixteenth and seventeenth centuries *froze one particular step* in this dialectical development, assigned to the temporarily defeated rivals a place outside science and so prevented their return.[15] Today science is on top because the show has been rigged in its favour and not because of any inherent excellence either of its methods or of its results.

(10) Modern science overtook its *scholarly* rivals in a competition that at least at some stage had the appearance of fairness and rationality.[16] The *primitive* views that were found during the expansion of the fifteenth and sixteenth centuries were never considered worthy of entering such a competition. They were simply pushed aside and replaced, first by Christianity, later by science. Their removal was not a result of research, but of a firm belief in the superiority of the white man and of all of his works. The excellence of science that is the basic creed of almost all institutional rationalists may therefore be nothing but a pious wish.

Just how far this pious wish is removed from reality has been shown by more recent research into older cultures and contemporary non-Western cultures and civilizations. We know now that Stone Age man possessed a fairly well-developed lunisolar astronomy that was used for practical purposes, tested in observatories and incorporated into social fables so that we have here an astronomy that is factually adequate, practically useful and socially relevant.[17] Considering that the average workday of Stone Age man was about four hours[18] we may conjecture that the astronomy became part of a philosophical world view of considerable sophistication.[19] Bits and pieces of such a world view that seem to have spread all over Europe as well as into India and China can be restored from later literary products.[20] They show an insight into the role of change that was superior by far to the

---

[15] Interestingly enough such freezing procedures did not occur in the arts; the discovery of central perspective was soon followed by mannerism and the revolt against strict and unrealistic rules.

[16] It is interesting to see that a more detailed examination of scientific episodes always shows them to be much less rational than everyone is inclined to think. <see> *Against Method*, chs. 6–12.

[17] <See> the reports in F. R. Hodson (ed.), *The Place of Astronomy in the Ancient World* (London, 1974).

[18] <See> the work of Marshall Sahlins.

[19] <See> Alexander Marshack, *The Roots of Civilisation* (New York: McGraw-Hill, 1972).

[20] On the work of the restorers <see> G. de Santillana and H. von Dechend *Hamlet's Mill: An Essay Investigating the Origins of Human Knowledge and its Transmission Through Myth* (Boston, 1969).

later assumption of 'eternal laws of nature'.[21] Speculation was combined with experiment and so we have now various theories and forms of healing that are better in diagnosis and therapy than the unwieldy, clumsy, though spectacular methods of modern scientific medicine. Scientific medicine seems successful because the point of comparison is its own average achievement. Choose a different and more realistic point of comparison and the success story turns into a story of dismal failure. And, mind you, the comparison would still be between a science that is being fed by billions of dollars of tax money and whose ideology is supported by the whole educational process and an opponent whose only strength (apart from sound theory and efficient practice) is the persistence of its followers. It is therefore very doubtful indeed whether a fair competition would today make science come out on top. Result: science cannot solve the problem of rationalism, it is itself part of the problem.

(11) Considerations such as these are a starting point for the *normative rationalist*. Normative rationalism points out that institutions and traditions have their ups and downs, that they are always capable of improvement, that even the most perfect institution may come off badly when compared with ideals different from those it tries to realize. Also, standards are expressed by thought-statements which can never be obtained from an analysis of what is, even if the object of the analysis should happen to be the fact that certain standards are used and held in high esteem. The domain of rationality is therefore separate from the domain of facts, traditions, institutions. Facts, traditions, institutions may be rational in the sense that they conform to the laws of this domain (so that it would be irrational to deny them, or to go against them), but they cannot give us the values and the standards that generate such judgements.

Now standards, or rules, are used not just because of the intellectual pleasure one may derive from their discussion. They are supposed to guide real actions, they are supposed to produce results in this world. 'Rational' procedures that run counter to sociological and psychological tendencies have not much chance of succeeding. Rules that demand actions contrary to physical laws are hopeless. 'Valid' standards that do not belong to any tradition might as well not exist. Even those standards that are examined in a purely intellectual fashion and in utter disregard of sociological and cosmological facts still form part of a tradition of intellectual debate (which may or may not be strong enough to influence, and perhaps even to change

---

[21] In Hesiod laws of nature are subjected to change and they are the result of a dynamic equilibrium between opposing forces (law of Zeus vs. the laws of the titans). Also, the basic principles of the universe have various aspects (live and generative, dead and passive) which come forth in different circumstances. All this is much closer to modern science than it was to the science of the nineteenth century: myth was in many respects closer to reality than the highly sophisticated scientific ideologies that replaced it.

primary traditions): there is no appeal to standards outside traditions.[22] The question is therefore not whether and how standards can influence traditions, the question is how certain traditions (intellectualistic traditions considering validity, truth, etc.) can influence other traditions: insofar as normative rationalism is supposed to have effects in this world it turns out to be a special version of institutional (cosmological) rationalism.

(12) With this we are back to the problem of section 7: there are many institutions – so why choose a particular one as a basis of rationality? The normative rationalist tries to solve the problem by appealing to a judge that is 'objective' and independent of traditions. He fails because a judge is effective only if he is institutionalized. Result: there is not one rationality, there are many and it is up to us to choose the one we like the best.

(13) For many thinkers such a result is intolerable. Relativism, they believe, opens the door to chaos and arbitrariness. The fear of chaos, the longing for a world in which one need not make fundamental decisions but can always count on advice, has made rationalists act like frightened children. 'What shall we do?', 'How shall we choose?', they cry when presented with a set of alternatives, assuming that the choice is not their own, but must be decided by standards that are (a) explicit and (b) not themselves subjected to a choice. Relativism, however, brings choice into everything – hence the aversion.

(14) The first objection against this assumption is that it gives standards a one-sided authority. Traditions, actions, decisions are measured by standards, standards are not measured by traditions (actions, decisions). This is, of course, the view of normative rationalists. But normative rationalism, insofar as it has an effect upon research, was found to be a special case of institutional rationalism and institutions, having all the same 'ontological status' <,> both influence and are influenced by, other institutions (traditions, etc.).

The second objection is a direct consequence of the reply to the first. Traditions, institutions influence behaviour not only via rules and standards that can be made explicit. When recording an observation, reacting to a smile, checking the results of a complicated calculation, we act 'automatically', without consulting explicit rules and without being able to say what rules were involved. Nor is it possible to avoid such behaviour. Assume we want to judge action A by standard S. We apply S to A and render our judgement. But the application must also be rational, so there must be standards S' which judge the pair (A, S) and so *ad infinitum* unless we admit

---

[22] This is a triviality for a Wittgensteinian (<see> my review of the *Philosophical Investigations* in *The Philosophical Review*, vol. 64, 1955).

that at some place we simply act without being able to provide the standards which make this action rational.[23]

This general observation is supported by concrete historical research.[24] Everyone agrees that Newton's celestial mechanics, Kepler's planetary theories, Maxwell's electrodynamics, the special and general theories of relativity are splendid achievements of rational thought. We also know the rules and standards that were popular when the theories were developed and other rules and standards which are today said to have led to them. None of these rules and standards would have permitted the theories to survive, and those that are more permissive are too weak to give any guidance whatsoever. One might conjecture that the correct rules will one fine day be found and that their discovery will reveal the rationality of all important episodes of science. The conjecture does not seem very plausible and, besides, the need to make it shows that the rationality of science in the sense of the assumption (of section 13) is nothing but a dream. Let us now collect <any> ideas that may give us a more realistic account.

(15) We have seen (section 11) that standards are not outside traditions, but are parts of them. We have also seen (section 12) that there are many traditions containing different sets of standards. Moreover, action, even complex action, can proceed and often does proceed without standards that are either explicit, or can be made explicit (section 14). And as extending a tradition into the future is always an open matter,[25] we may say that traditions not merely guide actions, but are constituted by them. Now actions which introduce new traditions lack even the proper starting point. They do not fit into any pre-existing pattern. In some cases they even clash with authoritative traditions and are therefore irrational in a very strong sense of the word.[26] Yet such 'homeless' and 'irrational' action, being joined by other homeless and irrational actions may coalesce into a new form of life which later generations regard as the very essence of reason.[27] A researcher, therefore, does not just *follow* rules and standards, he also *invents* them and in the course of his inventions often goes against what his own time calls 'reason', but what later on may turn out to be no more than the reason of the time.

(16) We can now return to the two forms of rationalism that were introduced at the beginning of the present short note. Naïve and sophisticated rationalists assume that each individual action, each individual piece of research must be subjected to their rules. The rules (standards) determine the structure of research in advance, they guarantee its objectivity,

---

[23] For details <see> the work of Michael Polanyi.
[24] For details concerning the assertions made in this paragraph <see> *Against Method*.
[25] <See> Wittgenstein on continuing <a> series of integers, etc.
[26] For an example, <see> *Against Method*, pp. 260–71.
[27] <See> *Against Method*, pp. 256f.

they guarantee that we are dealing with rational action. We have seen that every action and every piece of research may be regarded as a potential instance of the application of the rule, but it may also be regarded as a test case: we may permit the rule to guide research, we may permit research to suspend the rule. In making the latter decision we acknowledge that there are no rules apart from traditions (result of the criticism of normative rationalism), that traditions give not only explicit rules but also tendencies for action (result of the second and third objection in section 14), that the tendencies not only guide actions, but are also constituted by them, that actions may change them and introduce entirely new traditions. A researcher who deviates from the tradition in which he works does not rely on any clear insight into its limitations, for these limitations appear only when new traditions have arrived. He relies, rather, on a vague hope that he will find such traditions and will then be able to explain what at first seemed like madness and irrationality.[28] Not every researcher who does unusual things succeeds in this; nor is lack of success always to be ascribed to the irrationality of the ideas used; it is often due to the absence of historical circumstances which are needed if irrational actions are to coalesce into a new form of reason. Those thinkers, however, who do succeed show that scientists (philosophers, religious leaders) are inventors of theories, <of> instruments, as well as of entire forms of life which they introduce, bit by bit, against all rhyme and reason because rhyme and reason are often found only after one has moved a considerable distance without them.

Now a researcher who admits such possibilities (and I do not see how, considering the historical evidence, one can deny them) will not abolish any rules and standards. Rather, he will try to learn as many of them as possible, he will try to improve them, to make them more flexible, for on his ventures into the unknown he needs all the help he can get. He knows that every step he makes is a step into darkness. He may end up in obscurity and empty verbiage; but he may also find new canons of action and understanding. Even close attendance to the laws of the tradition in which he grew up and which dominates his surroundings does not increase the light on his path. His life may be safe, 'rational', secure, he may achieve fame among the public and earn the respect of his peers and yet all this, seen from an as-yet-undiscovered form of life, may be but a grandiose exercise in futility. So, here is a really interesting choice to be made. It is the choice between adherence to a predominant tradition and 'rationality' in the sense of this tradition on the one side, and the path of 'irrationality'

---

[28] The appearance of irrationality is often concealed by the fact that one continues to use the same *words* though their *meaning* is gradually bent in a new direction. For an example <see> p. 267, third paragraph, of *Against Method*.

which may, or may not, lead to a new and perhaps better form of life on the other. This choice confronts the scientist even at the most trite step of his research and it cannot be replaced by any appeal to standards. One might call the omnipresence of this choice the 'existential dimension' of research.[29] The fact that there is such an existential dimension to every single action we carry out shows that rationalism is not an agency that forms an otherwise chaotic material, but is itself material to be formed by personal decisions. The questions 'What shall we do? How shall we proceed? What rules shall we adopt? What standards are there to guide us?' however, are answered by saying: 'You are grown up now, children, and so you have to find your own way.'

[29] <See> Kierkegaard, *Concluding Unscientific Postscript*, as well as Polanyi's *Personal Knowledge: Towards a Post-Critical Philosophy* (Chicago: University of Chicago Press, 1958).

II

Democracy, élitism and scientific method

In the following brief note I want to examine a theoretical problem which some intellectuals have raised in connection with citizens' initiatives: how, i.e. on the basis of what standards, can the citizens of a democracy judge the institutions that surround them and the things they produce?

It is often assumed that they will have to choose 'rationally', which in turn is assumed to mean that they will have to choose in accordance with scientific standards. The difficulty with this assumption is that there are no unambiguous scientific standards. I begin my paper with an explanation of this difficulty.

I SCIENTIFIC RATIONALITY – A PHILOSOPHICAL DREAM

The nineteenth century was a period of vigorous methodological debate. In the natural sciences Newton played an important role. He had introduced a methodology and shown how it had aided him in his discoveries. According to this methodology science proceeds in a series of well-defined and clearly separated steps. First we find the facts (or 'phenomena', in Newton's terminology). Then we derive laws. Finally, we devise hypotheses for explaining the laws. Hypotheses and facts must be kept apart. It is not the imagination of the theoretician but the skill of the experimenter that determines what counts as a fact and how the facts are to be presented.

Newton's ideas had an important influence on the style of research and on the way in which research papers were written. Speculation was suspect, experimentation was the core of the sciences. In the course of the nineteenth century these tendencies were criticized by philosophers and scientists. Mill in his *Logic* established speculation as a legitimate method of research provided the hypotheses arising from it are used in accordance with certain rules. They must be *tested* by the derivation of predictions and the performance of experiments. Failure of a single test removes the corresponding hypothesis. A hypothesis that passes tests may be retained but it becomes part of knowledge only if it can be shown to be the only hypothesis capable of explaining the facts in its domain. Mill thought that Newton had succeeded in constructing a uniqueness proof for his own theory of gravitation.

Mill's essay *On Liberty* differs from his *Logic* both in style and in content. Mill is here interested in the development of individuality and talent. According to Mill people develop best in pluralistic societies that contain many ideas, traditions, forms of life. Such societies are also best suited for the improvement of knowledge. A plurality of views is preferable to a uniform intellectual climate 'on four different grounds'.[1] First, because a view one may have reason to reject may still be true. 'To deny this is to assume our own infallibility.' Secondly, because a problematic view 'may and very commonly does, contain a portion of truth; and since the general or prevailing opinion on any subject is rarely or never the whole truth, it is only by the collision of adverse opinions that the remainder of the truth has any chance of being supplied.' Thirdly, even a point of view that is wholly true but not contested 'will ... be held in the manner of a prejudice, with little comprehension or feeling of its rational grounds'. And, fourthly, one will not understand its meaning, accepting it will become 'a mere formal confession' unless a contrast with other ideas shows wherein this meaning consists.

The first two reasons are amply supported by the history of science. Ideas are often rejected before they can show their strength. Even in a fair competition one ideology, partly through accident, partly because greater attention is devoted to it, may assemble successes and overtake its rivals. This does not mean that the beaten rivals are without merit and have ceased to be capable of making a contribution. It only means that the special application made of them so far did not reveal their strong points or that they have temporarily run out of steam. They may return and cause the defeat of their defeaters. The philosophy of atomism is an excellent example. It was introduced (in the West) in antiquity with the purpose of 'saving' macrophenomena such as motion. It was overtaken by the dynamically more sophisticated philosophy of the Aristotelians, returned with the scientific revolution, was pushed back and almost annihilated during the nineteenth century, returned early in the twentieth century and is now again restricted by complementarity. Many facts that seemed to refute atomism on closer analysis turned out to support it. Or take the idea of the motion of the earth. It arose in antiquity, was then defeated by the powerful arguments of the Aristotelians, regarded as an 'incredibly ridiculous' view by Ptolemy, and yet staged a triumphant comeback in the seventeenth century. What is true of theories is true of methods: knowledge was founded on speculation and logic, then Aristotle introduced a more empirical procedure which was replaced by the mathematical methods of Descartes and Galileo, which in turn were

---

[1] J. S. Mill, 'On Liberty' quoted from M. Cohen (ed.), *The Philosophy of John Stuart Mill* (New York: Random House, 1961), pp. 245 f.

combined with a fairly radical empiricism by the members of the Copenhagen School. The lesson to be drawn from this historical sketch is that a setback for a theory, a point of view, an ideology must not be taken as a reason for eliminating it. The ideas and instruments that are needed to analyse complex facts in terms of one particular theory or research programme may arrive only long after another research programme has assembled impressive successes. *A science interested in finding truth must therefore retain all the ideas of mankind for possible use or, to put it differently: the history of ideas is an essential part of scientific method.*

Reasons 3 and 4 receive support from some very interesting and depressing phenomena that occur whenever an idea manages to become the centre of attention. The rise, success, triumph of a theory, point of view, philosophy almost always leads to a considerable decrease of rationality and understanding. When the view is first proposed it faces a hostile audience; excellent reasons must be given to gain it an even moderately fair hearing. The reasons are often disregarded, or laughed out of court, and unhappiness is the fate of the bold inventor. But if the reasons are understood and accepted, if there are some temporary successes, then interest may be increased and people may devote themselves to a close study of the basic ideas. Professional groups form and make these ideas sufficiently respectable to be represented at conferences and meetings. Even the die-hards of the *status quo* now feel an obligation to study the one or the other paper and to make a few grumbling remarks. There comes then a moment when the theory is no longer an esoteric discussion topic for advanced seminars but enters the process of education itself. Introductory texts are written, popularizations appear, examination questions contain problems to be solved in its terms. Scientists and philosophers from distant fields, trying to show their erudition, drop pregnant hints, and this often quite uninformed desire to be regarded as a well-informed person is taken as a further sign that the theory is an essential part of scientific knowledge. *But this increase in importance is not accompanied by better understanding.* Problematic aspects which were originally introduced with the help of arguments, in full awareness of their difficulties, are now turned into basic principles; doubtful points, having been generally accepted, become slogans, debates with opponents' rituals, for the opponents, now forced to use enemy terms to express their discontent seem to misuse them, or else to raise mere quibbles. Alternatives are still employed – but they only serve to underline the splendour of the new theory (cf. the role of 'classical physics' in arguments about the foundations of the quantum theory; or the role of 'inductivism' in the arguments of Popperians; cf. also the bogeymen liberals and Marxists have set up for easy refutation) in political debates. So finally we have success – but it is the success of a manoeuvre carried out in a void, not the success of a reasoned view overcoming real difficulties rather than obstacles set up for easy refutation. An empirical theory (as opposed to a philosophical theory) such as quantum mechanics or an empirical practice such as modern scientific medicine with its materialistic background can of course point to numerous results, but the trouble is that any view and any procedure that is developed and applied by intelligent beings has results; what we want to know is whose results are better and more important for those who receive them. This question cannot be answered, for after the triumph of a research programme there are no alternatives left to provide points of comparison.

Mill gives a clear and compelling description of these phenomena. Debates and reasoning, he writes, are features

> belonging to periods of transition, when old notions and feelings have been unsettled and no new doctrines have yet succeeded to their ascendancy. At such times people of any mental activity, having given up their old beliefs, and not feeling quite sure that those they still retain can stand unmodified, listen eagerly to new opinions. But this state of things is necessarily transitory: some particular body of doctrine in time rallies the majority around it, organises social institutions and modes of actions conformably to itself, education impresses this new creed upon the new generation *without the mental processes that have led to it* and by degrees it acquires the very same power of compression, so long exercised by the creeds of which it had taken place.[2]

(Note how clearly and simply Mill describes a phenomenon which Kuhn much later used to support a rather cumbersome theory of 'revolutions' and 'normal' periods.) An account of the alternatives replaced, of the process of replacement, of the arguments used in its course, of the strength of the old views and the weaknesses of the new, not a 'systematic' account, but a historical account of each stage of knowledge, can alleviate these drawbacks and increase the rationality of our allegiance to the views we regard as the most satisfactory.

Proliferation, for Mill, is therefore not only an expression of his liberal attitude, of his firm belief that people have a right to live as they see fit, of his faith that such a plurality of life styles and modes of thought will be of advantage to all; it *is* also *an essential part of any rational inquiry concerning the nature of things. 'Outmoded' views are kept alive both because they please some people and because the most advanced theories cannot be understood and examined without their help.*

Hegel criticized Newton's empiricism in a different way. He developed a theory of conceptual change that undercut all forms of empiricism and he also criticized Newton's notorious 'derivation' of the law of gravitation

---

[2] J. S. Mill 'Autobiography', quoted from Max Lerner (ed.), *Essential Works of John Stuart Mill* (New York: Knopf, 1965), p. 149, my emphasis.

from 'facts'. Hegel's ideas influenced dialectical materialists (up to Bohm) and the philosophy of the social sciences, but had little effect elsewhere.[3]

Both Hegel and Mill subjected methodological rules and standards of rationality to the on-going process of research (the on-going competition of alternatives in the case of Mill; the development of the absolute in the case of Hegel). For them knowledge was not a process guided by external rules and criteria, but a process that contained its own criteria, was subjected to them and influenced them in turn. In this respect Mill and Hegel were rather close to scientists who, prompted by developments such as discoveries in the field of electromagnetism, the debates about Darwin, the strife about the atomic hypothesis, the arguments used in connection with the '1847 biophysics programme', as it has been called (also known as the 'Helmholtz school'), and with the attempt to turn psychology into a science, produced a great variety of methodological views. For these scientists (scientific) reason was not an agency with clear and unambiguous demands that had to be obeyed, come what may, but rather a clever, inventive and many-sided guide through the vicissitudes of research. Maxwell, Boltzmann, Helmholtz, Hertz, Mach, Duhem all favoured a *methodological pluralism* that was guided by examples of past research. Each of these scientists was of course in favour of certain procedures and against others; but they all agreed that such personal preferences must not be turned into 'objective standards'. 'The best means of promoting the development of the sciences', writes Duhem after a vigorous diatribe against model building, 'is to permit each form of intellect to develop itself by following its own laws and realizing fully its type'.[4] 'I must admit', writes von Helmholtz, 'that so far I have retained the latter procedure [mathematical equations instead of models] and felt safe with it – but I would not like to raise general objections against a way which such excellent physicists ... have chosen.'[5] Boltzmann, after a survey of new methods in theoretical physics objects to regarding any one of them, old or new, as the only acceptable one.[6] 'The schematisms of *formal* logic and also of *inductive* logic are not of much use', writes Ernst Mach, 'for the intellectual situations are never repeated in exactly the same way. But the examples of great scientists are very instructive ...'[7] A special feature of Mach's philosophy is that science explores *all* aspects of knowledge, 'principles' as well as theories, 'foundations' as well as peripheral assumptions, local rules as well as the

---

[3] For details <see> Sect. 3 of 'Against Method', *Minnesota Studies in the Philosophy of Science, volume IV* (Minneapolis: University of Minnesota Press, 1970). [Now reprinted as chapter 4 of *PP2* (Ed.)].

[4] P. Duhem, *The Aim and Structure of Physical Theory* (New York: Atheneum, 1962), p. 99.

[5] H. L. von Helmholtz, 'Introduction' to Heinrich Hertz, *Die Prinzipien der Mechanik* (Leipzig: Johann Ambrosius Barth, 1894), pp. xxi f.

[6] L. Boltzmann, *Populäre Schriften* (Leipzig: Johann Ambrosius Barth, 1906), ch. 1. esp. p.10.

[7] E. Mach, *Erkenntnis and Irrtum* (Leipzig: Johann Ambrosius Barth, 1905), p. 200.

laws of logic;[8] it is an autonomous enterprise, not guided by ideas imposed without control from its own on-going research process.[9]

The fruitful and 'opportunistic' pluralism of these scientists survived in Einstein, who studied all of them,[10] and in Niels Bohr.[11] It played and still plays an important role in twentieth-century science. *It did not survive in philosophy.* For the 'Revolution in Philosophy' that introduced neopositivism put an end to the great age of proliferation. Philosophy of science became a special subject with special methods and an organon of its own: formal logic. The task was no longer to aid the sciences; the task was now to 'analyse' them, that is to 'translate' theories and procedures into a special language that lacked the elasticity and fruitful vagueness of the scientific language. The translation procedures soon got entangled with themselves so that even the analysis of science now takes second place to the tremendously important problem: how to get rid of this self-entanglement. Historians have shown that the image which these 'scientific' philosophers had of science was entirely fanciful: science does not agree with the rules and standards the positivists have provided for it – and any attempt to make it agree is bound to have disastrous consequences.[12]

Let me repeat that this clash between philosophical standards and scientific practice is not a new and revolutionary discovery. Many leading nineteenth-century scientists and some of their philosophical contemporaries denied the existence of a 'scientific method'; they pointed out that scientists proceed in the manner most appropriate to the subject under investigation, that new types of research often require new types of standards and that even the laws of formal logic are not exempt from revision. Twentieth-century philosophers of science slowly became aware of this situation, but they have not yet fully grasped it. Trying to retain the idea of general standards for all knowledge-creating and knowledge-improving activities, they either voided existing standards until they became mere slogans, without cognitive content, or else they developed

---

[8] <See> Boltzmann, *Populäre Schriften*, pp. 400ff.

[9] The difference between Mach and the philosophers is explained in *Science in a Free Society* (London: New Left Books, 1978), pp. 195ff. <See> also my essay 'In Defence of Aristotle', in G. Radnitzky and G. Andersson (eds.), *Progress and Rationality in Science* (Dordrecht: D. Reidel, 1978), Sect. 4.

[10] <See> Sect. 9 of P. Feyerabend, 'Zahar on Mach, Einstein and Modern Science', *British Journal for the Philosophy of Science*, <vol. 31, 1980, pp. 273–82>, which contains some relevant quotations on Einstein's self-professed 'opportunism'. <See> also *Against Method* (London: New Left Books, 1975), p. 18 fn. 6, as well as the introduction to my *Problems of Empiricism: Philosophical Papers, volume 2* (Cambridge: Cambridge University Press, <1981>).

[11] <See> my paper 'On a Recent Critique of Complementarity, <Parts I and II>', *Philosophy of Science*, vols. 35 and 36, 1968 and 1969, pp. 309–31 and 82–105.

[12] These matters are explained in detail in my *Against Method*, and in Pt 1 of my *Science in a Free Society* (London: New Left Books, 1978), as well as in the introduction to *Problems of Empiricism*.

them independently of science, as mere logical exercises. Not being aware of the voiding process in the first case and taking the connection with science for granted in the second, they could claim to have standards which are both substantial and relevant when in fact they are neither. A good example of the first alternative is *Lakatos*: his standards do have content but what they forbid is hardly worth fighting about: they only forbid scientists to *call* certain research programmes 'progressive' – everything else goes. Good examples of the second alternative are almost all modern theories of induction, probabilistic or otherwise. They 'clarify' abstract notions such as 'evidence', 'confirmed to a higher degree than', etc. and they take it for granted that the notions, being general and abstract, are bound to apply to all types of knowledge. They overlook that not all cultural entities have general properties of the kind used in the 'clarifications' and that scientific research is deficient in precisely this respect. This, then, is the end of the philosophy of science (and of a very popular idea of reason) as an aid to scientific progress.

2 POLITICAL CONSEQUENCES

The situation I have just described has interesting and troublesome political consequences.

It is generally agreed that a free society must not be left at the mercy of the institutions it contains – it must be able to supervise and control them. The citizens and the democratic councils that exercise the control must evaluate the achievements and the effects of the most powerful institutions. For example, they must evaluate the effects of science and take steps (withdrawal of financial support, reduction of scientific influence in elementary and high-school education, limitations and perhaps complete removal of academic freedom, and so on) if these effects turn out to be useless or harmful. To evaluate, the citizens need intellectual guides, they need standards. Now if standards for the evaluation of scientific research are research-immanent, if they change as research proceeds, and if their change can be controlled and understood only by those immersed in research, then a citizen who wants to judge science must either become a scientist himself, or he must defer to the advice of experts. A democratic control of science (and of other institutions) is then impossible.

This is indeed the conclusion that has been drawn by Michael Polanyi, one of the few twentieth-century scientist-philosophers to notice and assert the research-immanence of scientific standards. According to Polanyi there is no way in which outsiders can judge science. *Science knows best.* (Kuhn and Holton give the same answer, though in more muted terms.) Are we to concede that the end of the philosophy of science also means the end of a democratic control of science and scientists?

According to Lakatos, who devoted much attention to this problem, a democratic control of science (and of other institutions) is possible only if we have criteria, or standards which are *respectable* and which can be *separated* from scientific practice. The standards must be respectable, for we want to make a serious choice and not merely follow the whim of the moment. And they must be separable from scientific practice, for outsiders such as the common citizen must be able to learn, use and to apply them without becoming scientists. Lakatos' interest in general and situation-independent standards has philosophical as well as political motives.

To guarantee the respectability of his standards Lakatos connects them with science. To guarantee their separability he connects them only with special parts of science which can be expected to have certain features in common. The parts Lakatos chooses are achievements which everyone regards as important and path breaking. Newton's mechanics, Darwin's theory, the rise of the special and the general theory of relativity are achievements of this kind. Lakatos admits that science as a whole can be grasped by research-immanent standards only. The standards which judge science at its best, however, can be detached from scientific practice and understood independently of any mastery of it. Moreover, they have bite, for they can be turned against scientific developments that do not favourably compare with the events from which they were abstracted (according to Lakatos modern elementary particle physics, empirical sociology, psychoanalysis, astrology and parapsychology are such developments). Lakatos advises foundations, political bodies, and individual citizens to use the standards in their evaluation of the sciences and to withdraw money, political support, educational authority, etc., from developments that do not conform to them.

We must admit that Lakatos has seen a most important problem – but he has not solved it. The standards he proposes do not fit the parts of science he has chosen as a basis[13] and we never hear why a deviation from the standards should be regarded as a disadvantage and not as an improvement. Assume they change again in the twentieth century, for example in elementary particle physics. Why is this change not a further improvement? We get no answer. Thirdly, the standards permit us only to compare one part of science with another; they do not help us to judge science as a whole: Lakatos shares Polanyi's view of the role of science in society except that he chooses part of science as his measure while Polanyi chooses all of science. Hence, if Polanyi is a 'Stalinist' or ('élitist')[14] then so is Lakatos – except that he bases his Stalinism on a different and more narrow basis

[13] This is shown in *Against Method*, ch. 16

[14] For 'élitist' <see> I. Lakatos, *Philosophical Papers, volume 2* (Cambridge: Cambridge University Press, 1978), pp. 114 f. In talks and private conversations Lakatos used 'Stalinist' instead.

than Polanyi. The second point (imperviousness of the standards to a change in science) also shows that the sciences enter the arguments as window-dressing only. They are used for their propaganda value as long as they agree with a preconceived philosophical idea (the idea of content increase); they are dropped as soon as they move away from the idea. Lakatos' élitism is therefore the élitism of a *philosophical clique* that wants to intimidate people in exactly the same manner in which science has intimidated them so far.

With this we are back to our original problem: how is a citizen going to judge the suggestions issuing from the institutions that surround him, and how is he going to judge these institutions themselves? He needs criteria and standards – this is what our intellectuals tell us. What standards is he going to use?

3 INTELLECTUAL ÉLITISM V. DEMOCRATIC RELATIVISM

I think that in a free society the answer is obvious: *A citizen will use the standards of the tradition to which he belongs;* Hopi standards, if he is a Hopi; fundamentalist Protestant standards if he is a Fundamentalist; ancient Jewish standards, if he belongs to a group trying to revive ancient Jewish traditions; nor must we forget special groups which, realizing that they have special interests and ideas, try to act in accordance with them – I am thinking of the women's movements, gay liberation, ecological groups, and so on. Of course, all the groups need knowledge to apply the standards they use but the epistemic criteria which decide what is knowledge and what not are determined by the traditions themselves and not by outside agencies. It is also clear that people learn, and adopt ideas from other traditions, but this process again depends on the standards of the tradition that does the adopting. Finally, we must not overlook that today almost all traditions are part of larger units, they are parts of a city, a cluster of cities, a state, a confederation of states, and that they are constrained by the institutions and the laws of these units. It again depends on the traditions how they will deal with such constraints, for example, how they will use them to further their own interests. Some citizens of the state of California have used state laws to introduce ideas of *Genesis* into biology textbooks and to remove passages which presented evolution as a fact. Black Muslims became capitalists to increase their fiscal and spiritual independence. The citizens of Puerto Rico may soon succeed in obtaining their independence. Citizens' initiatives have stopped highways and nuclear reactors and have made possible the legal use of non-Western forms of medicine such as acupuncture. The freedom of a society increases as the restrictions imposed on its traditions are removed.

Note how this answer and the attitude that leads to it differ from the answer and the attitude of the intellectuals. Intellectuals raise questions such as: 'We want to judge the institutions of our societies; we need standards to do that. What are the correct standards? How do we find them?' In asking the question they assume that there is a problem; that any problem *they* perceive is a problem for everyone; and that they are the right people to solve such problems. *They simply take it for granted that their own traditions of standard-construction and standard rejection are the only traditions that count.* We have an élitist answer – we do not have a democratic answer.

But the medicine men of central African tribes have no trouble forming an opinion about 'scientific' medicine: they let Western physicians explain their business, they consider the matter, they accept certain forms of treatment and reject others. They have standards and they know how to use them – even in unusual circumstances. Women, trusting more in the regenerative powers of Nature than in the male presumption that sickness is a malfunction that can and must be corrected by scientific tinkering, have found their own ways of dealing with a great variety of disturbances. And so on. The fact that intellectuals have mental blind-spots certainly does not mean that everybody has.

Alien traditions, it is replied, may have answers where Western intellectuals have problems – but these answers cannot be taken seriously. Science, technology, medicine and other institutions that have been developed in the West are better than all alternatives, because they have *results*. This is why their problems are real problems which must be taken seriously by everybody. This reply, which many philosophers accept without hesitation, has no rational basis. The sciences, it is said, are uniformly better than all alternatives – but where is the evidence to support this claim? Where, for example, are the control groups which show the uniform (and not only the occasional) superiority of Western scientific medicine over the medicine of the *Nei Ching*? Or over Hopi medicine? Such control groups need patients that have been treated in the Hopi manner, or in the Chinese manner using Hopi experts and experts in traditional Chinese medicine (rather than Western physicians who have learned the one or the other exotic trick and now already regard themselves as 'experts' in the alien art), but the requested procedures are often against the law and are at any rate frowned upon and sabotaged by the medical societies. Secondly, the reply assumes what is to be shown, viz. standards that make the results of science worthwhile. But a mystic who can leave his material body and meet God Himself will hardly be impressed by the fact that thousands of people, using billions of dollars of tax money, succeeded in putting two wrapped-up bodies on a hot and dried-out stone, the moon, and he will deplore the decrease and almost complete destruction of man's spiritual abilities that is a result of the materialistic–scientific climate of our times. One may of course ridicule such an observation – but one cannot remove it by using the argument from scientific success. Difference in standards and values plays an even greater role in medicine: Western 'scientific' medicine aims at the smooth functioning of the body-machine no matter what its feelings or its aesthetic appearance; other forms of medicine are interested in feelings, intuitive abilities, special achievements<,> prophecy, Shamanism, that cannot be measured in materialistic terms.

Another objection to a democratic relativism is that we live in a scientific age and have to adapt to it. The reply is, first, that this is not true – science is by no means omnipresent – and, secondly, that even its omnipresence cannot be regarded as an argument for acceptance: if a country is invaded by locusts then it is useful to study their habits, but it would be quite unreasonable to turn them into national deities.

The élitism attacked, so goes a further argument, does no harm, for today everybody can in principle become a member of the élite: everybody can become a scientist, a politician, a Great Thinker, even the President of a University. However, one can become a member of the élite only if one adopts its ideology and its habits: equality, including equality of women and 'racial' equality, does not mean equality of traditions; *it means equality of access to one particular tradition* – the tradition of the White Man. White Liberals supporting the demand for equality have opened the promised land – but it is a promised land built according to their own specifications, filled with their own favourite playthings and accessible only in accordance with their particular requirements (cf. the importance of 'intelligence' tests for access to all sorts of activities).[15]

The most effective move against a democratic relativism – for we can hardly speak of an argument – is emotional blackmail or, more correctly, slander. For example, many critics raise the spectre of racism, Auschwitz, terrorism, chaos. But democratic relativism denies the right of traditions to impose their form of life on others, and it therefore recommends the protection of traditions from interference from outside. Hopi medicine will be protected from Western medico-fascism just as Jews will be protected from the political fascism of the Anti-Semites. Nor is the fear of chaos justified: the traditions whose independence we want to protect are usually much stricter towards their members than is the protecting mechanism towards the protected traditions. The belief that the institutions of a free society should protect the individual and not traditions, is closely connected with the liberal belief that individuals can exist and have properties worth protecting independently of all traditions. The belief is correct to a certain extent: a foetus already has individuality, it reacts towards its surroundings,

---

[15] Only few radicals have noticed this restriction. Thus women liberationists fight for the right of women to participate in male manias and only few women are critical of these manias themselves.

it contains the possibilities for a rich and rewarding life. What is not correct is the assumption that preservation of these possibilities is a basic value never to be overruled. Not even liberals make this assumption a basis of their creed (not all liberals are pacifists). Besides, a foetus is not a fully fledged human being; it needs a tradition to become that, and so traditions do become the prime elements of society. Of course there will be cases when the state rightfully interferes even with the internal business of the traditions it contains (to prevent the spreading of infectious diseases, for example), for as with every rule the rules of democratic relativism have exceptions. The point is that in a democracy the nature and the placement of the exceptions is determined by specially elected groups of citizens and not by experts, and that these groups will choose a democratic relativism as the basis on which the exceptions are imposed. The problem of education, however (people may remain in execrable traditions because they do not know better, hence we need a uniform and universal education) provides an argument *against* the *status quo*, not *for* it: hardly any adherent of science, of scientific medicine, of rational procedures has chosen this form of life from among a variety of alternatives, the scientific point of view was imposed by 'education', not chosen[16] and the groups who want to leave the fold and return to more traditional forms of life do so in full knowledge of the splendours they are leaving behind: they have savoured the bouquet of scientific rationalism and have found it wanting. We see that neither arguments nor moral pressures can *remove* the democratic relativism that was proposed at the beginning of the present section. And there are many arguments *in its favour*.[17]

The first argument, which I have already mentioned, is one of *rights*. *People have the right to live as they see fit.* If there exists a tradition that has religious reasons for rejecting certain forms of medical treatment (some tribes in Central Africa do not want to be X-rayed because they do not want their internal organs exposed to view), then no institution should be permitted to force it to accept these forms. Conversely, if there is a tradition that uses treatment contrary to the ideas of Western medicine, then no institution should be permitted to force it to reject these forms, or to put them at a disadvantage (no health insurance, for example, or no paid sick leave). Science or rationalism, in this view, are instruments put at the disposal of the people *to be used by them as they see fit;* they are *not* necessary

---

[16] According to Kant, enlightenment occurs when people leave the stage of self-inflicted immaturity. On the basis of this definition we can describe the development since the eighteenth century by saying that immaturity *vis-à-vis* the churches was replaced by immaturity *vis-à-vis* science and rationalism. Enlightenment is as distant today as it was in the sixteenth century.

[17] A more detailed version of these arguments is found in pt. 2, ch. 3 of my *Erkenntnis für freie Menschen*, 2nd, improved edition (Frankfurt: Suhrkamp, 1980).

conditions of rationality, or citizenship, or life. Scientists are salesmen of ideas and gadgets, they are not judges of Truth and Falsehood. Nor are they High Priests of Right Living. I have already said that there are exceptions to this rule, as there are to any rule. The point is that in a democracy these exceptions are dealt with by democratic councils, and that these councils take democratic relativism as their starting-point.

A second argument in favour of a democratic relativism is closely connected with Mill's arguments for proliferation. A society that contains many traditions side by side has much better means of judging each single tradition than a monistic society. It enhances both the quality of the traditions, and the maturity of its citizens. We can learn a lot from 'primitive' tribes about care for the aged, treatment of 'criminal' elements, treatment of (behavioural) aberrations. We can observe the advantages of direct knowledge over an 'objective' account that approaches its object in a standardized and severely alienating way. For Montaigne and his followers in the Enlightenment a study of savage cultures provided not only valuable contributions to the understanding of man, it also put a mirror in front of 'civilized' man and exposed his shortcomings and vices. Today a look at the lives of independent *women* can show us the barbarism that characterizes so much of our *man*-made societies. Democratic relativism makes these contrasts stand out and so enables everybody to learn from them in her or his own way.

A third argument follows directly from the second. Scientific views are not only incomplete, in that they omit important phenomena, they are often erroneous at the very centre of their competence. Routine arguments and routine procedures are based on assumptions which are inaccessible to the research of the time and often turn out to be either false or nonsensical. Examples are the views on space, time, and reality in eighteenth- and nineteenth-century physics and astronomy, the materialism of most medical researchers today, the crude empiricism that guided much of seventeenth- and eighteenth-century science and influenced even the debates about the Darwinian theory. These views are essential parts of important research traditions, and yet only few practitioners know them and can talk about them intelligently. And yet scientists show increased belligerence when the views are attacked, and they throw their whole authority behind ideas they can neither formulate nor defend: who does not remember how vigorously some scientists have defended a rather naïve form of empiricism without even being able to say what facts are or why anyone should take them seriously. The lesson to be learned from this phenomenon is that *fundamental debates between traditions are debates between laymen which can and should be settled by no higher authority than again the authority of laymen, i.e. democratic councils.*

The situation just described is important in cases where the belief in the soundness of a view or enterprise has led to institutional (and not only intellectual) measures against alternatives. 'Scientific' medicine is a good example. It is not a monolithic entity, it contains many departments, schools, ideas, procedures and a good deal of dissension. However, there are some widespread assumptions which influence research at many points but are never subjected to criticism. One is the assumption that illnesses are due to material disturbances which can be localized and identified as to their chemico-physiological nature, and that the proper treatment consists in removing them either by drugs or by surgery (including complex methods of surgery such as laser surgery). The question is whether the problems of scientific medicine – and there are many – have something to do with this assumption, or whether their origin lies elsewhere. Everybody knows that death rates in hospitals go down when doctors go on strike. Is this due to the incompetence of the physicians, or does it show a basic fault in the theoretical structure that guides their actions? Everybody knows that cancer research absorbs huge amounts of money and has few results.[18] Is this due to the fact that cancer researchers are mainly interested in theory, not in healing, or does it indicate a basic fault of the theories used? We do not know. To find out we must first make the basic assumptions (materialism, for example) visible, and then examine them in a more direct manner. To examine them in a more direct manner we have to compare the results of scientific medicine with those of forms of medicine based on entirely different principles. Democratic relativism permits and protects the practice of such different forms of medicine.[19] It makes the needed comparisons possible. *Democratic relativism*, therefore, not only *supports a right, it is also a most useful research instrument for any tradition that accepts it.*

4 FIRST STEPS

Democratic relativism, then, is a Fine Thing, but how are we going to introduce it? And how are we going to keep the various traditions in place and prevent them from overwhelming each other by force, as Western conquerors once overwhelmed old cultures? The answer is that the necessary institutions *already exist:* almost all traditions are part of societies with a firmly entrenched protective machinery. The question is therefore not how to *construct* such a machinery, but how to *loosen it up* and how to detach it from the traditions now using it exclusively for their own purposes. For example, the question is how to separate state and science. The answer to *this* question is that the methods employed cannot be

---

[18] <See> Daniel Greenberg, 'The 'War on Cancer': Official Fiction and Health Facts', *Science and Government Reports*, vol. 4, 1 December 1974.

[19] The objection that people must be protected cannot be raised at this stage: after all, we do not yet know, prior to the comparison, what they must be protected from. It may well turn out that we must protect them from mutilation by 'scientific' body-plumbers.

discussed independently of the tradition that wants to achieve equality, and the situation it finds itself in. The democratic relativism I have discussed will not be imposed *from above*, by a gang of radical intellectuals, it will be realized *from within*, by those who *want* to become independent, and in the manner *they* find most suitable (if they are a lazy bunch they will move *very* slowly, and with long periods of rest in between their political inter- ventions). What counts are not intellectual schemes, but the wishes of those who want change. Or, to use a catchy slogan: *citizens' initiatives instead of philosophy!* For details the reader is referred to my *Science in a Free Society* and to the much improved German version, *Erkenntnis für Freie Menschen*.[20]

[20] See notes 9 and 17.

12

Appendix:
The works of Paul Feyerabend

Eric Oberheim

Paul Feyerabend did not keep a list of his publications. Shortly before he died, he put together from memory a short, incomplete, and sometimes erroneous list of his publications, which served as the starting point for this bibliography.

First and foremost, I wish to express my gratitude to Grazia Borrini-Feyerabend. Without her valuable assistance and generous hospitality this bibliography would not have been possible. Daniel Sirtes, Paul Hoyningen-Huene, Marko Toivanen, K. Hofmann, and John Preston also generously provided time and information for which I am grateful. I would like to thank a number of colleagues who kindly proof-read the foreign-language entries, most especially Nicola Teubner. I also thank the editors of the *Journal for General Philosophy of Science* for the many corrections that were made. Last but not least, I would like to thank B. Uhlemann at the Philosophical Archive of the University of Konstanz for her support.

Feyerabend's books are, for the most part, collages of his articles. I have tried to show this by cross-referencing. Dates followed by capital letters refer to books, while dates followed by lower-case letters refer to articles.

BOOKS

(1944 Nov.–1955 May 9, inclusive, 1944 Nov.–1948 June 19, bulk) *Tagebuch.* Diary written in German shorthand. Now deposited in the Feyerabend Archive at the University of Konstanz (Reference PF 3–3–1), 141 sheets and 31 attachments.

(1951) *Zur Theorie der Basissätze.* (PhD Dissertation) (Universitäts Bibliothek Wien), 149 pp.

(1955) *Humanities in Austria: A Report on Postwar Developments.* Translated by Ingeborg Satinger, Vilma Anderl and Elvin Schlecht (Washington DC: Library of Congress Reference Department).

(1961) *Knowledge Without Foundations: Two Lectures Delivered on the Nellie Heldt Lecture Fund* (Oberlin, OH: Oberlin College), 74 pp. (reprinted in this volume).

(1971) *I problemi dell' empirismo.* Translated by Anna Maria Siolo (Milano: Lampugnani Negri Editore), 192 pp.

(1975A) *Against Method. Outline of an Anarchistic Theory of Knowledge.* (London: New Left Books and Atlantic Highlands, NJ: Humanities Press), 339 pp. Reprinted 1978 (London and New York: Verso).

(1976) *Wider den Methodenzwang. Skizze einer anarchistischen Erkenntnistheorie.* Translated by Hermann Vetter. English text revised and expanded by Feyerabend (Frankfurt am Main: Suhrkamp), 443 pp.

(1977) *In strijd met de methode. Aanzet tot een anarchistische kennistheorie.* Translated from English by Hein Kray (Boom: Meppel), 375 pp.

(1977) *Contra o método: Esboça de una teoria anárquica da teoria do contiecimento.* Translated by Octanny S. da Mota and Leonidas Hegenberg (Rio de Janeiro: Livraria Francisco Alves), 487 pp.

(1977) *Ned med metodologin! Skiss till en anarkistisk kunskapsteori.* Translated by Thomas Brante (Zenit: Rabén and Sjögren), 326 pp.

(1979) *Contro il metodo. Abbozzo di una teoria anarchica della conscenza Paul K. Feyerabend.* Translated by Libero Sosio. With a foreword by Giulio Giorello (Milano: Feltrinelli), viii + 262 pp. Reprinted 1984.

(1979) *Contre la méthode. Esquisse d'une théorie anarchiste de la connaissance.* Translated from the English by Baudouin Jurdant and Agnès Schlumberger (Paris: Editions du Seuil), 350 pp. Reprinted 1988.

(1981) *Contra el método: Esquema de una teoria anarquista del conocimiento.* Translated by Francisco Hernán (Barcelona: Ariel), 207 pp. Reprinted 1987.

(1981) *Tratado contra el método.* Translated by Diego Ribes (Madrid: Tecnos), xvii + 319 pp.

(1981) *Hoho eno chosen: Kagakuteki sozo to chi no anakizumu* (Tokyo: Shin'yosha), 13 + 438 pp.

(1983) *Wider den Methodenzwang.* Translated by Hermann Vetter. Revised by Paul K. Feyerabend (Frankfurt am Main: Suhrkamp): 423 pp. Reprinted 1986 and 1987.

(1984) *Contra el método.* Translated by Francisco Hernán (Orbis: Esplugas de Llobregat), 192 pp.

(1988) *Against Method.* Revised edition (London and New York: Verso), viii + 296 pp.

(1989) *Yönteme Hayir: Bir Anarsist Bilgi Kuraminin Ana Hatlari.* Translated by Ahmet Inam (Istanbul: Ara Yayincilik Baski), 325 pp.

(1993) *Against Method.* Third edition (London and New York: Verso), xiv + 279 pp.

(1994) Translated into Chinese by Changzhong Zhou (Shanghai: Shanghai Translation Publishing House), 269 pp.

(1978A) *Der wissenschaftstheoretische Realismus und die Autorität der Wissenschaften: Ausgewählte Schriften, Band 1* (Braunschweig, Wiesbaden: Vieweg), viii + 367 pp. Reprinted 1980.

(1978B) *Science in a Free Society* (London: New Left Books), 221 pp.

(1979) *Erkenntnis für freie Menschen* (Frankfurt am Main: Suhrkamp), 272 pp.

(1980) *Erkenntnis für freie Menschen.* Revised edition (Frankfurt am Main: Suhrkamp), 300 pp. Reprinted 1981, 1987, 1995.

(1981) *La scienza in una società libera* (Milano: Feltrinelli), 234 pp. Reprinted 1982.

(1982) *Jiyujin no tameno chi: Kagakuron no kaitai e* (Tokyo: Shin'yosha), ix + 333 pp.

(1982) *Science in a Free Society* (London: Verso), 221 pp. Reprinted 1983.

(1982) *La ciencia en una sociedad libre.* Translated by Alberto Elena (Madrid: Siglo XXI de Espanã), 272 pp.

(1987) *Özgür bir Toplumda Bicim.* Translated by Ahmet Kardam (Istanbul: Basiminden çevrilmistir), 265 pp.

(1990) Translated from the English (Verso, 1982) into Chinese by Yazhu Lan (Shanghai: Shanghai Translation Publishing House), 257 pp.

(1981A) *Realism, Rationalism and Scientific Method: Philosophical Papers Volume 1* (Cambridge: Cambridge University Press), xiv + 353 pp. Reprinted 1983, 1985.

(1981B) *Problems of Empiricism: Philosophical Papers Volume 2* (Cambridge: Cambridge University Press), xii + 255 pp. Reprinted 1983, 1985.

(1981C) *Probleme des Empirismus. Schriften zur Theorie der Erklärung, der Quantentheorie und der Wissenschaftsgeschichte. Ausgewählte Schriften, Band 2* (Braunschweig, Wiesbaden: Vieweg), xiv + 472 pp.

(1984) *Wissenschaft als Kunst* (Frankfurt am Main: Suhrkamp), 169 pp. Reprinted 1985, 1993.

  (1984) *Scienza come arte.* Translated by L. Sosio. With an introduction by Marcello Pera and supplemented by Paul K. Feyerabend (Roma: Laterza), xxxiii + 199 pp.

(1987A) *Farewell to Reason* (London and New York: Verso), 327 pp. Reprinted 1988.

  (1989) *Adieu à la raison.* Translated by Baudouin Jurdant (Paris: Les Editions du Seuil), 373 pp.

  (1989) *Irrwege der Vernunft.* Translated by Jürgen Blasius (Frankfurt am Main: Suhrkamp), 471 pp. Reprinted 1990.

  (1990) *Addio alla ragione.* Translated by Marcello D'Agostino (Roma: Armando Armando), 320 pp.

  (1992) *Risei yo saraba* (Tokyo: Hoseidaigakushuppankyoku), 34 + 399 pp.

(1989A) *Dialogo sul metodo.* Translated by Roberta Corvi (Roma and Bari: Laterza e Figli), 175 pp.

  (1989) *Dialogo sobre el metodo.* Translated by José Casas (Roma and Bari: Laterza e Figli), 165 pp.

  (1991) *Diálogo sobre o método.* Translated by António Guerreiro (Lisboa: Prescença), 137 pp.

  (1991) *Three Dialogues on Knowledge* (Oxford: Blackwell), 167 pp.

  (1991) *Dialoghi sulla conoscenza.* Translated by Roberta Corvi (Roma and Bari: Laterza e Figli), 120 pp.

  (1991) *Dialogos sobre el conocimiento.* Translated by Jerónima García Bonafé (Roma and Bari: Laterza e Figli), 120 pp.

  (1992) *Über Erkenntnis. Zwei Dialoge.* Translated by Ilse Griem and Hans Günther Holl (Frankfurt am Main and New York: Campus), 210 pp. The English text appears as the second and third Dialogue in *Three Dialogues on Knowledge* (1991).

  (1993) *Over kennis. Twee dialogen.* Translated by Maarten van der Marei (Kampen, Kok Agora: DNB/Pelckmans, Kapellen), 126 pp.

  (1993) Translated into Japanese by Youichirou Murakami (Tokyo: Tuttle-Mori), 304 pp.

  (1995) *Die Torheit der Philosophen: Dialoge über die Erkenntnis.* Translated by Henning Thies (Hamburg: Junius), 154 pp.

  (1995) *Bilgi Üzerine üç söylesi.* Translated by C. Güzel (Beyoglu, Istanbul: Metis Yoyinlari), 187 pp.

  (in press) Translated into French by Baudouin Jurdant (Paris: Les Editions du Seuil).

(1994) *Ammazzando il tempo. Un'autobiografia.* Translated by Alessandro de Lachenal (Roma and Bari: Laterza e Figli), 219 pp.

(1995) *Killing Time: The Autobiography of Paul Feyerabend* (Chicago and London: University of Chicago Press), 192 pp.

(1995) *Zeitverschwendung.* Translated by Joachim Jung (Frankfurt am Main: Suhrkamp), 250 pp. Reprinted 1995.

(1995) *Matando el tiempo: Autobiografía.* Translated by Fabián Chueca (Madrid: Editorial Debate), xxviii + 177 pp.

(in press) *Tuer le temps.* Translated by Baudouin Jurdant (Paris: Les Editions du Seuil).

(1996) *Thesen zum Anarchismus, und andere Texte* (Berlin: Karin Kramer Verlag), 240 pp.

(1998) *Widerstreit und Harmonie. Trentiner Vorlesungen.* P. Engelmann (ed.). Translated by S. Rodl (Vienna: Passagen Philosophie), 200 pp.

(in hand) *Knowledge, Science and Relativism: Philosophical Papers Volume 3.* John Preston (ed.) (Cambridge: Cambridge University Press).

(in press) *The Conquest of Abundance.* Bert Terpstra (Editor).

Preface: based on *Killing Time*, pp. 163–4 and 179–80.

A Note on the Editing (by Bert Terpstra).

Part 1   Conquest of Abundance

     Introduction

        Ch.1   Achilles' Passionate Conjecture

        Ch.2   Xenophanes

        Ch.3   Parmenides and the Logic of Being

        Interlude   On the Ambiguity of Interpretations

        Ch.4   Brunelleschi and the Invention of Perspective

Part 2   Articles

        Ch.1   Realism and the Historicity of Knowledge

        Ch.2   Quantum Theory and Our View of the World

        Ch.3   Excerpt from 'Realism'

        Ch.4   What Reality?

        Ch.5   Aristotle

        Ch.6   Art as a Product of Nature as a Work of Art

        Ch.7   Ethics as a Measure of Scientific Truth

        Ch.8   Universals as Tyrants and as Mediators

        Ch.8   Intellectuals and the Facts of Life

        Ch.9   Concerning an Appeal for Philosophy

Postscript and Acknowledgements (by Grazia Borrini-Feyerabend).

PAUL FEYERABEND AS EDITOR

(1966) *Mind, Matter, and Method: Essays in Philosophy and Science in Honor of Herbert Feigl.* Paul K. Feyerabend and Grover Maxwell (eds.) (Minneapolis: University of Minnesota Press), 524 pp.

Feyerabend's contributions:

'Herbert Feigl: A Biographical Sketch', pp. 3–13

'On the Possibility of a Perpetuum Mobile of the Second Kind', pp. 409–12.

(1976) *Essays in Memory of Imre Lakatos.* R. S. Cohen, Paul K. Feyerabend and M. W. Wartofsky (eds.) (Dordrecht and Boston: D. Reidel) vii + 767 pp.
Feyerabend's contributions:
'Preface', by Paul K. Feyerabend, Robert S. Cohen and Marx W. Wartofsky, p. vi,
'On the Critique of Scientific Reason', pp. 109–43.

(1982) *Von der Verantwortung des Wissens: Positionen der neueren Philosophie der Wissenschaft.* P. Good and P. K. Feyerabend (eds.) (Frankfurt am Main: Suhrkamp).
Feyerabend's contributions:
'Hermeneutischer Probleme der praktischen Vernunft'.

(1983) *Wissenschaft und Tradition.* Paul Feyerabend and Christian Thomas (eds.) Eidgenössische Technische Hochschule Zürich (Zürich: Verlag der Fachvereine), xi + 303 pp.
Feyerabend's contributions:
'Einleitung', pp. 1–3
'Briefwechsel', P. K. Feyerabend and P. R. Ruhstaller, pp. 19–20
'Die Frage, ob die Theologie eine Wissenschaft sei, ist weder interessant noch gehaltvoll', pp. 125–8
'Diskussion', pp. 133–8
'Institutionelle Methoden der Wahrheitsgewinnung und Wahrheitsfestlegung in den Wissenschaften', pp. 139–48
'Auszüge aus Dokumenten zum <Prozess Galilei>', pp. 177–81
'Der Galileiprozess – einige unzeitgemässe Betrachtungen', pp. 183–92
'Auszug aus dem Urteil des Distriktrichters gegen das Land Arkansas vom 5. Januar 1982', translated and summarized by P. K. Feyerabend, pp. 227–30
'Diskussion', pp. 237–8
'Keine Erkenntnis ohne Kunst', pp. 289–95
'Diskussion', p. 303.

(1984) *Kunst und Wissenschaft.* Paul Feyerabend and Christian Thomas (eds.). Eidgenössische Technische Hochschule Zürich (Zürich: Verlag der Fachvereine), x + 250 pp.
Feyerabend's contributions:
'Einleitung', pp. 1–5
'Ganzheits- und Aggregatauffassungen, illustriert am Beispiel der Kontinuität und der Bewegung', pp. 151–8
'Kreativität – Grundlage der Wissenschaften und der Künste oder leeres Gerede?', pp. 187–99
'Quantitativer und qualitativer Fortschritt in Kunst, Philosophie und Wissenschaft', pp. 217–30.

(1985) *Grenzprobleme der Wissenschaften.* Paul Feyerabend and Christian Thomas (eds.). Eidgenössische Technische Hochschule Zürich (Zürich: Verlag der Fachvereine), xviii + 429 pp.
Feyerabend's contributions:
'Einleitung', pp. 1–3

'Begrifflichkeit: Grundlage von Wissenschaft. Diskussionsbeitrag', pp. 123–5
'Diskussionsbeitrag zu Liebe und Tod', pp. 179–81
'Die Rolle von Fachleuten in einer freien Gesellschaft', pp. 325–33
'Was heisst das, wissenschaftlich zu sein?', pp. 385–97.

(1986) *Nutzniesser und Betroffene von Wissenschaften.* Paul Feyerabend and Christian Thomas (eds.). Eidgenössische Technische Hochschule Zürich (Zürich: Verlag der Fachvereine), x + 241 pp.
Feyerabend's contributions:
'Einleitung', pp. 1–6
'Diskussionsbeitrag zu: Experten in der Politik', pp. 121–2
'Diskussionsbeitrag zu: Biologen als Benutzer von Physik und Chemie', pp. 131–2
'Diskussionsbeitrag zu: Studenten zwischen Aussenseitertum und Berufsstolz', pp. 161–5
'Irrtümer und Betrügereien in den Wissenschaften. Diskussionsbeitrag', pp. 213–18
'Demokratie, Elitarismus und wissenschaftliche Methode', Translated from the English by Dr. Margrit Soland, pp. 219–41.

(1987) *Leben mit den 'Acht Todsünden der zivilisierten Menschheit'? Eine aktuelle Diskussion an der ETH Zürich zu den Thesen von Konrad Lorenz.* Paul Feyerabend and Christian Thomas (eds.). Eidgenössische Technische Hochschule Zürich (Zürich: Verlag der Fachvereine), xxv + 416 pp.
Feyerabend's contributions:
'Einleitung', pp. 1–8
'Bemerkungen zu Varela: Ein Weg entsteht im Unterwegssein', pp. 325–9
Single contributions in 'Zusammenfassung der Diskussion', pp. 353–68.

(1992) *Philosophie, Psychoanalyse, Emigration: Festschrift für Kurt Rudolf Fischer zum 70. Geburtstag.* Peter Muhr, Paul Feyerabend and Cornelia Wegeler (eds.) (Wien: WUV–Universitätsverlag), 488 pp.

ARTICLES, REVIEWS, INTERVIEWS,
PUBLISHED LETTERS AND A FILM

(1947) 'Der Begriff der Anschaulichkeit in der modernen Physik', *Veröffentlichungen des Österreichischen College*, Vienna.
(1948) Feyerabend played a small role in the film *Der Prozess* directed by G.W. Pabst, and starring Ernst Deutsch.
(1952a) 'Neuere Probleme der philosophischen Logik', *Wissenschaft und Weltbild: Monatsschrift für alle Gebiete der Forschung* (5) 9, pp. 315–19.
(1952b) 'Bemerkungen zu "Interpretation and Preciseness"', *Semantiske Problemer*, Oslo, pp. 77–84.
(1954a) 'Wittgenstein und die Philosophie, I', *Ibid.* (7) 5–6, pp. 212–20.
(1954b) 'Wittgenstein und die Philosophie, II', *Ibid.* 7–8, pp. 283–7.
(1954c) 'Physik und Ontologie', *Ibid.* 11–12, pp. 464–76.

(1954d) 'Ludwig Wittgenstein', *Merkur. Deutsche Zeitschrift für europäisches Denken* (8) 11, pp. 1021–38.

(1954e) 'Determinismus und Quantenmechanik', *Wiener Zeitschrift für Philosophie, Psychologie, Pädagogik* (5) 2, pp. 89–111.

(1954f) Review of Rudolf Carnap's *Einführung in die symbolische Logik* mit besonderer Berücksichtigung ihrer Anwendungen, *Internationale mathematische Nachrichten* (9) 35–6, p. 67.

(1955a) Review of Ludwig Wittgenstein's *Philosophical Investigations, The Philosophical Review* (64) 3, pp. 449–83 (see also 1981B, 99–130, 1966b, 1967o).

(1955b) 'Carnaps Theorie der Interpretation theoretischer Systeme', *Theoria* (21) pp. 55–62.

(1956a) 'A Note on the Paradox of Analysis', *Philosophical Studies* (7) 6, pp. 92–6.

(1956b) 'Eine Bemerkung zum Neumannschen Beweis', *Zeitschrift für Physik* (145) 4, pp. 421–3.

(1956c) Review of *Kant's Metaphysics and Theory of Science* by Gottfried Martin, *British Journal for the Philosophy of Science* (7) 27, pp. 260–3.

(1957a) Review of *Foundations of Quantum-Mechanics: A Study in Continuity and Symmetry* by A. Landé, *British Journal for the Philosophy of Science* (7) 28, pp. 354–7.

(1957b) 'Zur Quantentheorie der Messung', *Zeitschrift für Physik* (148) 5, pp. 551–9 (see also 1981A, 207–18, 1957c).

(1957c) 'On the Quantum-Theory of Measurement', *Observation and Interpretation: A Symposium of Philosophers and Physicists. Proceedings of the Ninth Symposium of the Colston Research Society.* S. Körner and M. H. L. Pryce (eds.) (London and New York: Butterworths Scientific Publications), pp. 121–30. Reprinted 1962 (New York and London: Dover Publications) (see also 1981A, 207–18, 1957b).

(1957d) 'Discussion remarks', *Ibid.* 21, pp. 48–9, 52, 53, 112–13, 139, 146–7, 172, 182, 183, 185.

(1957e) 'Die analytische Philosophie und das Paradox der Analyse', *Kant-Studien* (49) 3, pp. 238–44 (see also 1956a).

(1958a) 'Naturphilosophie', *Das Fischer Lexikon. Enzyklopädie des Wissens, Band 11: Philosophie* (Frankfurt am Main: Fischer Bücherei), pp. 203–27.

(1958b) 'An Attempt at a Realistic Interpretation of Experience', *Proceedings of the Aristotelian Society* (58) pp. 143–70 (see also 1981A, 17–36).

(1958c) 'Complementarity', *Ibid. Supplement* (32) pp. 75–104.

(1958d) 'Reichenbach's Interpretation of Quantum-Mechanics', *Philosophical Studies* (9) 4, pp. 49–59 (see also 1981A, 236–46, 1975a).

(1958e) Review of *Mathematical Foundations of Quantum Mechanics* by John von Neumann, translated by R. T. Beyer, *British Journal for the Philosophy of Science* (8) 32, pp. 343–7.

(1959a) Review of *The Direction of Time* by Hans Reichenbach, *British Journal for the Philosophy of Science* (9) 36, pp. 336–7.

(1960a) 'On the Interpretation of Scientific Theories', *Proceedings of the 12th International Congress of Philosophy, Venice, 1958, Volume 5* (Firenze: Sansoni Editore), pp. 151–9 (see also 1981A, 37–43, 1972d, 1973b).

(1960b) 'O Interpretacij Relacyj Nieokreslonosci', *Studia Filozoficzne* (19) pp. 21–78.

(1960c) 'Professor Landé on the Reduction of the Wave Packet', *American Journal of Physics* (28) 5, pp. 507–8.

234   ERIC OBERHEIM

(1960d) Review of *Patterns of Discovery. An Inquiry into the Conceptual Foundations of Science* by N. R. Hanson, *The Philosophical Review* (69) 2, pp. 247–52.

(1960e) 'Professor Bohm's Philosophy of Nature', Review of '*Causality and Chance in Modern Physics* by David Bohm', *British Journal for the Philosophy of Science* (10) 40, pp. 321–38 (see also 1981A, 219–35, 1970l).

(1960f) 'Das Problem der Existenz theoretischer Entitäten', *Probleme der Wissenschaftstheorie. Festschrift für Viktor Kraft.* E. Topitsch (ed.) (Wien: Springer Verlag), pp. 35–72 (see also 1981A, 40–73, and this volume for an English translation).

(1960g) 'Ludwig Wittgenstein', *Ludwig Wittgenstein, Schriften, Beiheft.* With a contribution by Ingeborg Bachmann u.a (Frankfurt am Main: Suhrkamp), pp. 30–47.

(1961a) 'Comments on Hanson's "Is There a Logic of Scientific Discovery?"', *Current Issues in the Philosophy of Science. Symposia of Scientists and Philosophers. Proceedings of Section L of the American Association for the Advancement of Science, 1959.* Herbert Feigl and Grover Maxwell (eds.) (New York: Holt, Rinehart and Winston), pp. 35–9.

(1961b) 'Comments on Sellars' "The Language of Theories"', *Ibid.*, pp. 82–3.

(1961c) 'Comments on Grünbaum's "Law and Convention in Physical Theory"', *Ibid.*, pp. 155–61.

(1961d) 'Comments on Barker's "The Role of Simplicity in Explanation"', *Ibid.*, pp. 278–80.

(1961e) 'Niels Bohr's Interpretation of the Quantum Theory', *Ibid.*, pp. 371–90.

(1961f) 'Rejoinder to Hanson', *Ibid.*, pp. 398–400.

(1961g) 'Comments on Hill's "Quantum Physics and Relativity Theory"', *Ibid.*, pp. 441–3.

(1961h) 'Méthodologie', *Les grands courants de la pensée mondiale contemporaine.* IIe Partie. Les tendances principales. M. F. Sciacca (ed.) (Milan: Marzorati-Editeur), pp. 871–99.

(1961i) 'Philosophie de la nature', *Ibid.*, pp. 901–27.

(1961j) Review of *Metascientific Queries* and *Causality* by Mario Bunge, *The Philosophical Review* (70) pp. 396–405.

(1961k) Review of *An Introduction to the Logic of the Sciences* by Rom Harré, *British Journal for the Philosophy of Science* (12) 47, pp. 245–50.

(1962a) 'Explanation, Reduction and Empiricism', *Scientific Explanation, Space and Time. Minnesota Studies in the Philosophy of Science Volume 3.* Herbert Feigl and Grover Maxwell (eds.) (Minneapolis: University of Minnesota Press), pp. 28–97. Reprinted 1966 (see also 1981A, 44–96, 1989d).

(1962b) 'Problems of Microphysics', *Frontiers of Science and Philosophy: University of Pittsburgh Series in the Philosophy of Science Volume I.* R. G. Colodny (ed.) (Pittsburgh: University of Pittsburgh Press), pp. 189–283. Reprinted 1964 (London: Allen and Unwin) (see also 1962c, 1967d).

(1962c) 'Problems of Microphysics', *Voice of America Broadcasts*, pp. 1–8 (see also 1962b, 1967d).

(1963a) 'Determinism and Indeterminism', *Harper's Encyclopedia of Science.* James D. Newman (ed.) (New York: Harper and Row), p. 320.

(1963b) 'Fictions (in science)', *Ibid.*, p. 444.

(1963c) 'Phenomenalism', *Ibid.*, p. 907.

(1963d) 'How to be a Good Empiricist; A Plea for Tolerance in Matters Epistemological', *Philosophy of Science: The Delaware Seminar Vol II.* B. Baumrin (ed.) (New York: Interscience Publishers), pp. 3–39 (see also 1968e, 1970h, 1970l, 1972c, 1976i, 1979c, 1980i, 1982g, and this volume).

(1963e) 'Professor Hartmann's Philosophy of Nature', Review of *Philosophie der Natur,* by Nicolai Hartmann, *Ratio* (5) 1, pp. 81–94.

(1963f) 'Mental Events and the Brain', *The Journal of Philosophy* (60) 11, pp. 295–6 (see also 1970h, 1981h, 1990d, 1993n).

(1963g) 'Materialism and the Mind–Body Problem', *The Review of Metaphysics* (17) 1, pp. 49–66 (see also 1981A, 161–75, 1967k, 1969e, 1970i, 1993l).

(1963h) Review of *Erkenntnislehre* by V. Kraft, *British Journal for the Philosophy of Science* (13) 52, pp. 319–23.

(1963i) Review of '*The Orgins of Science* by E. H. Hutten', *Isis* (54), no.178, pp. 487–8.

(1963j) 'Über konservative Züge in den Wissenschaften, insbesondere in der Quantentheorie, und ihre Beseitigung', *Club Voltaire: Jahrbuch für kritische Aufklärung, Band 1.* Gerhard Szczesny (ed.) (München: Szczesny Verlag), pp. 280–93. Reprinted 1964. Reprinted 1969 (Reinbeck bei Hamburg: Rowohlt Taschenbuch Verlag).

(1964a) Review of *The Concept of the Positron. A Philosophical Analysis* by N. R. Hanson, *The Philosophical Review* (73) 2, pp. 264–6.

(1964b) Review of *Scientific Change* edited by A. C. Crombie, *British Journal for the Philosophy of Science* (15) 59, pp. 244–54.

(1964c) 'A Note on the Problem of Induction', *The Journal of Philosophy* (61) 12, pp. 349–53 (see also 1981A, 203–6).

(1964d) 'Albert Einstein', *RIAS.*

(1964e) 'Realism and Instrumentalism: Comments on the Logic of Factual Support', *The Critical Approach to Science and Philosophy: In Honor of Karl R. Popper.* Mario Bunge (ed.) (London and New York: The Free Press of Glencoe), pp. 280–308 (see also 1981A, 176–202).

(1964f) 'Arbeitsgemeinschaften, stetige und unstetige Veränderungen in der Natur', *Das ist Alpbach.* Wolfgang Pfaundler and Barbara Coudenhoue-Kalergi (eds.) (Wien and München: Verlag Herold), p. 46.

(1965a) 'Reply to Criticism. Comments on Smart, Sellars and Putnam', *Proceedings of the Boston Colloquium for the Philosophy of Science 1962–64: In Honor of Philipp Frank. Boston Studies in the Philosophy of Science, Volume II.* Robert S. Cohen and Marx W. Wartofsky (eds.) (New York: Humanities Press), pp. 223–61.

(1965b) 'On the "Meaning" of Scientific Terms', *The Journal of Philosophy* (62) 10, pp. 266–74 (see also 1981A, 97–103, 1973c).

(1965c) Review of *Conjectures and Refutations. The Growth of Scientific Knowledge* by Karl R. Popper, *Isis* (56) p. 88.

(1965d) 'Wissenschaftstheorie', *Handwörterbuch der Sozialwissenschaften: Zugleich Neuauflage des Handwörterbuchs der Staatswissenschaften, Volume 12.* Erwin v. Beckerath et al (eds.) (Stuttgart: Gustav Fischer; Tübingen: J. C. B. Mohr; Göttingen: Vandenhoeck and Ruprecht), pp. 331–6.

(1965e) 'Problems of Empiricism', *Beyond the Edge of Certainty. Essays in Contemporary Science and Philosophy. University of Pittsburgh Series in the Philosophy of Science, Volume*

2. R. G. Colodny (ed.) (Englewood Cliffs, NJ: Prentice-Hall), pp. 145–260. Reprinted 1983 (Lanham, MD: University Press of America).

(1965f) 'Eigenart und Wandlungen physikalischer Erkenntnis', *Physikalische Blätter* (21) pp. 197–203.

(1966a) 'Dialectical Materialism and the Quantum Theory', *Slavic Review* (25) 3, pp. 414–17.

(1966b) 'Wittgenstein's *Philosophical Investigations*', Wittgenstein, *The Philosophical Investigations: A Collection of Critical Essays*. George Pitcher (ed.) (Garden City, N.Y.: Anchor Books, Doubleday and Co.), pp. 104–50. Reprinted 1968 (Notre Dame, IN: University of Notre Dame Press) (see also 1981B, 99–130, 1955a, 1967o).

(1966c) Review of *The Structure of Science* by Ernest Nagel, *British Journal for the Philosophy of Science* (17) 3, pp. 237–49 (see also 1981B, 52–64).

(1966d) Review of '*Die sprachphilosophischen und ontologischen Grundlagen im Spätwerk Ludwig Wittgensteins* by E. K. Specht', *The Philosophical Quarterly* (16) 62, pp. 79–80.

(1966e) 'Bemerkungen zur Verwendung nicht-klassischer Logiken in der Quantentheorie', *Deskription, Analytizität und Existenz. Forschungsgespräche des internationalen Forschungszentrums für Grundfragen der Wissenschaften Salzburg. Drittes und viertes Forschungsgespräch.* Paul Weingartner (ed.) (Salzburg and München: Anton Pustet), pp. 351–9 (see also 1978A, 113–20).

(1966f) 'Ist die Unterscheidung zwischen analytischen und synthetischen Sätzen überflüssig?' by H. Delius, W. Leinfellner, R. Haller, P. Feyerabend, H. Feigl, P. Weingartner, J. Hintikka, B. Juhos, H. Batuhan, A. Wilhelmy, R. Wohlgenannt, *Ibid.*, pp. 361–70.

(1966g) 'Analytische Sätze und wissenschaftliche Sprachsysteme' by P. Feyerabend, W. Leinfellner, J. Hintikka, B. Thum, B. Juhos, H. Feigl, P. Weingartner, *Ibid.*, pp. 370–4.

(1966h) 'Konventionalismus und Analytizität' by J. Hintikka, B. Juhos, P. Weingartner, P. Feyerabend, H. Feigl, A. Wilhelmy, *Ibid.*, pp. 374–7.

(1966i) 'Analytizität der Farbsätze' by P. Feyerabend, W. Leinfellner, H. Delius, B. Juhos, R. Wohlgenannt, P. Weingartner, R. Haller, J. Hintikka, K. Acham, H. Feigl, *Ibid.*, pp. 377–81.

(1966j) 'Verschiedene Bedeutungen von "Analytisch" ' by J. Hintikka, P. Feyerabend, W. Stegmüller, W. Leinfellner, B. Juhos, W. Essler, *Ibid.*, pp. 384–7.

(1966k) '"Analytisch-wahr" als "wahr in allen möglichen Welten" by W. Leinfellner, J. Hintikka, W. Stegmüller, A. Wilhelmy, P. Feyerabend, B. Juhos, *Ibid.*, pp. 387–90.

(1966l) 'Nicht-klassische Logik und ihre Anwendung' by W. Leinfellner, P. Feyerabend, B. Juhos, A. Wilhelmy, *Ibid.*, pp. 397–9.

(1966m) ' "Cogito ergo sum" by J. Hintikka', P. Weingartner, W. Leinfellner, B. Juhos, A. Wilhelmy, R. Haller, P. Feyerabend, R. Wohlgenannt, *Ibid.*, pp. 399–406.

(1967a) 'Bemerkungen zur Geschichte und Systematik des Empirismus', *Grundfragen der Wissenschaften und ihre Wurzeln in der Metaphysik, Fünftes Forschungsgespräch.* Paul Weingartner (ed.) (Salzburg and München: Universitätsverlag Anton Pustet), pp. 136–80 (see also 1978A, 249–92).

(1967b) 'The Theatre as an Instrument of the Criticism of Ideologies: Notes on Ionesco', *Inquiry* (10) 3, pp. 298–312 (see also 1978A, 139–52, 1967c).

(1967c) 'Theater als Ideologiekritik. Bemerkungen zu Ionesco', *Die Philosophie und die Wissenschaften. Simon Moser zum 65. Geburtstag.* Ernst Oldemeyer (ed.) (Meisenheim am Glan: Verlag Anton Hain), pp. 400–12 (see also 1978A, 139–52, 1967b).

(1967d) 'Problems of Microphysics', *Philosophy of Science Today.* Sidney Morgenbesser (ed.) (New York and London: Basic Books), pp. 136–47 (see also 1962b, 1962c).

(1967e) Review of *Law and Psychology in Conflict* by J. Marshall, *Inquiry* (10) 1, pp. 114–20.

(1967f) Review of *Philosophic Foundations of Quantum Mechanics* by Hans Reichenbach, *British Journal for the Philosophy of Science* (17) 4, pp. 326–8.

(1967g) 'Boltzmann, Ludwig', *The Encyclopedia of Philosophy Volume I.* Paul Edwards (ed.) (New York and London: Macmillan), pp. 334–7.

(1967h) 'Heisenberg, Werner', *Ibid., Volume 3*, pp. 466–7.

(1967i) 'Planck, Max', *Ibid.* Volume 6, pp. 312–14.

(1967j) 'Schrödinger, Erwin', *Ibid., Volume 7*, pp. 332–3.

(1967k) 'The Mind-Body Problem', *Continuum* (5) 1, pp. 35–49 (see also 1981A, 161–75, 1963g, 1969e, 1970i, 1993l).

(1967l) 'Von den Vorteilen des kalten Krieges', *Club Voltaire: Jahrbuch für kritische Aufklärung, Bd. 3.* Gerhard Szczesny (ed.) (München: Szczesny Verlag), pp. 38–45, 372. Reprinted 1969 (Reinbeck bei Hamburg: Rowohlt Taschenbuch Verlag).

(1967m) 'On the Improvement of the Sciences and the Arts, and the Possible Identity of the Two', *Proceedings of the Boston Colloquium for the Philosophy of Science, 1964–66: In Memory of Norwood Russell Hanson. Boston Studies in the Philosophy of Science, Volume III.* Robert S. Cohen and Marx W. Wartofsky (eds.) (Dordrecht: D. Reidel), pp. 387–415.

(1967n) 'Homage to N. R. Hanson', *Ibid.,* pp. xxii–iii.

(1967o) 'Wittgenstein's Philosophical Investigations', *Ludwig Wittgenstein. The Man and his Philosophy.* K.T. Fann (ed.) (Sussex: Humanities Press), pp. 214–50. Reprinted 1978 (see also 1981B, 1955a, 1966b).

(1967p) 'Alltagssprache, Wissenschaftssprache, Philosophensprache' by P. Feyerabend, G. Frey, R. Haller, B. Juhos, B. Thum, R. Wohlgenannt, E. Weinzierl, *Grundfragen der Wissenschaften und ihre Wurzeln in der Metaphysik, Fünftes Forschungsgespräch.* Paul Weingartner (ed.) (Salzburg and München: Universitätsverlag Anton Pustet), pp. 191–8.

(1967q) 'Begriff und Idee' by P. Lorenzen, P. Feyerabend, G. Frey, R. Wohlgenannt, P. Weingartner, K. Acham, B. Juhos, *Ibid.,* pp. 198–204.

(1967r) 'Wahrheitsdefinition und Existenzbegriff' by B. Juhos, P. Lorenzen, R. Haller, W. Del-Negro, P. Weingartner, C. Lejewski, P. Feyerabend, R. Wohlgenannt, *Ibid.,* pp. 204–15.

(1967s) 'Naturwissenschaft und Ontologie' by P. Feyerabend, B. Juhos, G. Frey, W. Del-Negro, B. Thum, R. Haller, P. Lorenzen, P. Weingartner, *Ibid.,* pp. 219–27.

(1967t) 'Metaphysische Voraussetzungen der Erfahrungswissenschaften' by B. Thum, B. Juhos, P. Feyerabend, G. Frey, P. Lorenzen, W. Del-Negro, *Ibid..,* pp. 227–32.

(1967u) 'Die Revision wissenschaftlicher Theorien' by H. R. Post, P. Feyerabend, P. Lorenzen, G. Frey, B. Thum, P. Weingartner, B. Juhos, *Ibid.*, pp. 232–42.

(1967v) 'Philosophy of Science', *Encyclopedia for Law and the Social Sciences.*

(1968a) 'Science, Freedom, and the Good Life', *The Philosophical Forum* (1) 2, pp. 127–35.

(1968b) 'A Note on Two "Problems" of Induction', *British Journal for the Philosophy of Science* (19) pp. 251–3.

(1968c) 'On a Recent Critique of Complementarity: Part I', *Philosophy of Science* (35) 4, pp. 309–31.

(1968d) 'Busoni's Aesthetics', Angel Records.

(1968e) 'How to be a Good Empiricist; A Plea for Tolerance in Matters Epistemological', *The Philosophy of Science. Oxford Readings in Philosophy.* P. H. Nidditch (ed.) (Oxford: Oxford University Press), pp. 12–39. Reprinted 1971, 1974, 1977 (see also 1963d, 1970h, 1970l, 1972c, 1976i, 1979c, 1980i, 1982g).

(1968f) 'Dear Russ. Homage to Norwood Russell Hanson', *Philosophy of Science* (35) pp. xxii–xxiii.

(1969a) 'Outline of a Pluralistic Theory of Knowledge and Action', *Planning for Diversity and Choice.* S. Anderson (ed.) (Cambridge, MA: MIT Press), pp. 275–84 (see also 1971b, and this volume).

(1969b) 'On a Recent Critique of Complementarity: Part II', *Philosophy of Science* (36) 1, pp. 82–105.

(1969c) 'Science without Experience', *The Journal of Philosophy* (66) 22, pp. 791–4 (see also 1981A, 132–5, 1972b).

(1969d) 'Linguistic Arguments and Scientific Method', *Telos* (2) pp. 43–63 (see also 1978A, 121–38, 1981A, 146–60, 1969/70).

(1969e) 'Materialism and the Mind–Body Problem', *Modern Materialism: Readings on Mind-Body Identity.* John O'Connor (ed.) (New York: Harcourt, Brace, and World), pp. 82–98 (see also 1981A, 161–75, 1963g, 1967k, 1970i, 1993l).

(1969/70) 'Questioni linguistiche e metodo scientifico', *Aut Aut* (19) pp. 114–15, 10–31 (see also 1978A, 121–38, 1981A, 146–60, 1969d).

(1970a) 'In Defence of Classical Physics', *Studies in History and Philosophy of Science* (1) 1, pp. 59–85.

(1970b) 'Experts in a Free Society', *The Critic* (29) pp. 58–69 (see also 1976d, and this volume).

(1970c) 'Consolations for the Specialist', *Criticism and the Growth of Knowledge. Proceedings of the International Colloquium in the Philosophy of Science, London, 1965, Volume 4.* Imre Lakatos and Alan Musgrave (eds.) (London and New York: Cambridge University Press), pp. 197–230. Reprinted 1973, 1974, 1976, 1977, 1978, 1979, 1980 (see also 1981B, 131–61, 1974i).

(1970d) 'Against Method: Outline of an Anarchistic Theory of Knowledge', *Analysis of Theories and Methods of Physics and Psychology: Minnesota Studies in the Philosophy of Science, Volume 4.* M. Radner and S. Winokur (eds.) (Minneapolis: University of Minnesota Press), pp. 17–130 (see also 1975A, 1973d, 1974g).

(1970e) 'Discussion at the Conference on Correspondence Rules' by P. Achinstein, I. G. Barbour, M. Brodbeck, R. Buek, J. W. Cornman, W. Craig, H. Feigl, P. Feyerabend, A. Grünbaum, N. R. Hanson, C. G. Hempel, M. B. Hesse,

E. L. Hill, E. McMullin, G. Maxwell, W. W. Rozeboom, W. C. Salmon, C. M. Van Vliet and F. M. Williams, *Ibid.*, pp. 220–59.

(1970f) 'Classical Empiricism', *The Methodological Heritage of Newton.* R. E. Butts and J. W. Davis (eds.) (Oxford and Toronto: Basil Blackwell), pp. 150–70 (see also 1981B, 34–51).

(1970g) 'Philosophy of Science: A Subject with a Great Past', *Historical and Philosophical Perspectives of Science. Minnesota Studies in the Philosophy of Science, Volume 5.* Roger H. Stuewer (ed.) (Minneapolis: University of Minnesota Press), pp. 172–83 (see also 1974h, and this volume).

(1970h) 'Wie wird man ein braver Empirist? Ein Aufruf zur Toleranz in der Erkenntnistheorie' translated by I. Meyer-Abich, *Erkenntnisprobleme der Naturwissenschaften. Texte zur Einführung in die Philosophie der Wissenschaft.* L. Krüger (ed.) (Köln and Berlin: Kiepenheuer und Witsch), pp. 302–35 (see also 1963d, 1968e, 1970h, 1970l, 1972c, 1976i, 1979c, 1980i, 1982g).

(1970i) 'Comment: "Mental events and the brain"', *The Mind/Brain Identity Theory. Controversies in Philosophy.* C. V. Borst (ed.) (London: Macmillan), pp. 140–1. Reprinted 1973, 1975, 1977 (see also 1963f, 1981h, 1990d, 1993n).

(1970j) 'Materialism and the Mind-Body Problem', *Ibid.*, pp. 142–56 (see also 1981A, 161–75, 1963g, 1967k, 1969e, 1993l).

(1970k) 'Problems of Empiricism, Part II', *The Nature and Function of Scientific Theories. Essays in Contemporary Science and Philosophy, University of Pittsburgh Series in the Philosophy of Science, Volume 4.* Robert G. Colodny (ed.) (Pittsburgh: University of Pittsburgh Press), pp. 275–353.

(1970l) 'How to be a Good Empiricist; A Plea for Tolerance in Matters Epistemological', *Readings in the Philosophy of Science.* B. A. Brody (ed.) (Englewood Cliffs, NJ: Prentice-Hall), pp. 319–42. 2nd edition 1989, 104–23 (see also 1963d, 1968e, 1970h, 1972c, 1976i, 1979c, 1980i, 1982g).

(1970m) 'Professor Bohm's Philosophy of Nature', *Physical Reality, Philosophical Essays on Twentieth-Century Physics.* Stephen Toulmin (ed.) (New York, Evanston and London: Harper and Row), pp. 173–96 (see also 1960e).

(1971a) 'Über einen neuen Liberalismus', *Radio München.*

(1971b) 'Umriß einer pluralistischen Theorie des Wissens und Handelns', *Die Zukunft der menschlichen Umwelt.* St Anderson (ed.) (Freiburg: Rombach), pp. 259–68 (see also 1969a).

(1971c) 'Theorie, Theoriebildung', *Lexikon der Pädagogik Bd 4* (Freiburg: Herder Verlag) (New edition), pp. 226–77.

(1971d) 'Wissenschaftstheorie', *Ibid.*, pp. 382–3.

(1972a) 'Von der beschränkten Gültigkeit methodologischer Regeln', *Neue Hefte für Philosophie* (2–3) pp. 124–71 (see also 1978A, 205–48, and this volume for an English translation).

(1972b) 'Science without Experience', *Challenges to Empiricism.* Harold Morick (ed.) (Belmont, CA: Wadsworth), pp. 160–4. Reprinted 1980 (London: Methuen) (see also 1981A, 132–5, 1969c).

(1972c) 'How to be a Good Empiricist; A Plea for Tolerance in Matters Epistemological', *Ibid.*, pp. 164–93 (see also 1963d, 1968e, 1970h, 1970kl, 1976i, 1979c, 1980i, 1982g).

(1972d) 'Über die Interpretation wissenschaftlicher Theorien', *Theorie und Realität:*

Ausgewählte Aufsätze zur Wissenschaftslehre der Sozialwissenschaften. Die Einheit der Gesellschaftswissenschaften, Band II. Hans Albert (ed.) (Tübingen: J. C. B. Mohr), pp. 59–66 (see also 1981A, 37–43, 1959a, 1973b).

(1972e) 'Philosophy of Science', *Encyclopedia for Education*, Freiberg.

(1972f) 'Scientific Laws', *Ibid.*,

(1973a) 'Die Wissenschaftstheorie – eine bisher unbekannte Form des Irrsinns?', *Natur und Geschichte. X. Kongress für Philosophie, Kiel 8.-12. Oktober 1972.* Kurt Hübner and Albert Menne (eds.) (Hamburg: Felix Meiner Verlag), pp. 88–124 (see also 1978A, 293–338).

(1973b) 'On the Interpretation of Scientific Theories', *Theories and Observation in Science.* Richard E. Grandy (ed.) (Englewood Cliffs, NJ: Prentice-Hall), pp. 147–53 (see also 1981A, 37–43, 1959a, 1972d).

(1973c) 'On the "Meaning' of Scientific Terms"', *Ibid.*, pp. 176–83 (see also 1965b).

(1973d) *Contro il metodo* (Milano: Lampugnani Nigri), 197 pp. (see also 1975A, 1970d, 1974g).

(1974a) 'Zahar on Einstein', *British Journal for the Philosophy of Science* (25) pp. 25–8.

(1974b) 'Thesen zum Anarchismus', *Unter dem Pflaster liegt der Strand* (1), Hans Peter Duerr (ed.) (Berlin: Karin Kramer Verlag), pp. 127–33.

(1974c) 'Die Wissenschaften in einer freien Gesellschaft', *Wissenschaftskrise und Wissenschaftskritik. Philosophie Aktuell, Band 1.* Walther Ch. Zimmerli (ed.) (Basel and Stuttgart: Schwabe and Co.), pp. 107–19 (see also 1978B).

(1974d) 'Brief an Doktor Zimmerli', *Ibid.*, pp. 136–7.

(1974e) 'Diskussion' by Jürgen Klüver, Heinz Kimmerle, Kurt Weisshaupt, Paul K. Feyerabend and Walter Ch. Zimmerli, *Ibid.*, pp. 131–40.

(1974f) Review of *Objective Knowledge. An Evolutionary Approach* by Karl R. Popper, *Inquiry* (17) 4, pp. 475–507 (see also 1981B, 168–201).

(1974g) *Contra el método. Esquema de una teoría anarquista del conocimiento.* Translated by Fransisco Hernán (Col. Ariel Quincenal, 85) (Ariel: Esplugas de Llobregat), 207 pp. (see also 1975A, 1970d, 1973d).

(1974h) 'Filosofia de la ciencia: Una materia con un gran pasado', translated by R. Beneyto, *Teorema* (4) pp. 11–27 (see also 1970f).

(1974i) 'Kuhns Struktur wissenschaftlicher Revolutionen – ein Trostbüchlein für Spezialisten?', *Kritik und Erkenntnisfortschritt. Abhandlungen des Internationalen Kolloquiums über die Philosophie der Wissenschaft, London 1965, Bd. 4.* Imre Lakatos and Alan Musgrave (eds.). Translated by P. K. Feyerabend and A. Szabó (Braunschweig: Friedr. Vieweg and Sohn), pp. 191–222 (see also 1981B, 131–61, 1970c).

(1974j) 'Imre Lakatos †', *Ibid.*, pp. vii–viii.

(1974k) 'Machamer on Galileo', *Studies in History and Philosophy of Science* (5) 3, pp. 297–304.

(1975a) 'Reichenbach's Interpretation of Quantum-Mechanics', *The Logico-Algebraic Approach to Quantum Mechanics. Volume I, Historical Evolution.* Clifford A. Hooker (ed.) (Dordrecht: D. Reidel), pp. 109–21 (see also 1981A, 236–46, 1958d).

(1975b) 'How to Defend Society Against Science', *Radical Philosophy* (11) pp. 3–8 (see also 1980f, 1981c, 1984c, 1988g, and this volume).

(1975c) 'Über einen neuen Versuch, die Vernunft zu retten', translated by Ilse Spinner-Offterdinger and Helmut F. Spinner, *Kölner Zeitschrift für Soziologie und*

Sozialpsychologie. Sonderheft (18) Wissenschaftssoziologie: Studien und Materialien. Nico Stehr and René König (eds.) (Opladen: Westdeutscher Verlag), pp. 479–514.

(1975d) 'Wie die Philosophie das Denken verhunzt und der Film es fördert', *Unter dem Pflaster liegt der Strand* (2). Hans Peter Duerr (ed.) (Berlin: Karin Kramer Verlag), pp. 224–37 (see also 1975e).

(1975e) 'Let's Make more Movies', *The Owl of Minerva: Philosophers on Philosophy*, Ch. J. Bontempo and S. J. Odell (eds.) (New York: McGraw-Hill), pp. 201–10 (see also 1975d, and this volume).

(1975f) ' "Science." The Myth and its Role in Society', *Inquiry* (18) 2, pp. 167–81.

(1975g) 'Imre Lakatos', *British Journal for the Philosophy of Science* (26) 1, pp. 1–18.

(1976a) 'Logic, Literacy, and Professor Gellner', *British Journal for the Philosophy of Science* (27) 4, pp. 381–91 (see also 1978B revised).

(1976b) 'Appendix, a Reply to Agassi', *Philosophia: Philosophical Quarterly of Israel* (6) 1, pp. 177–89 (see also 1978B, 1988f).

(1976c) 'Memorial Minutes (2)', *PSA 1974: Proceedings of the 1974 Biennial Meeting Philosophy of Science Association. Boston Studies in the Philosophy of Science, Volume 32.* Robert S. Cohen, C. A. Hooker, Alex C. Michalos and James W. Van Evra (eds.) (Dordrecht: D. Reidel), pp. xii–xiii.

(1976d) 'Experten in einer freien Gesellschaft', *Unter dem Pflaster liegt der Strand* (3) Hans Peter Duerr (ed.) (Berlin: Karin Kramer Verlag), pp. 11–35 (see also 1970b).

(1976e) 'Über die Methode. Ein Dialog', *Ibid.*, pp. 99–160 (see also 1989A, 1979d, 1981f).

(1976f) 'The Rationality of Science', *Can Theories be Refuted? Essays on the Duhem-Quine Thesis.* Sandra G. Harding (ed.) (Dordrecht and Boston: D. Reidel), pp. 289–315. Reprinted from 'Against Method' (1970d).

(1976g) 'On the Critique of Scientific Reason', *Method and Appraisal in the Physical Sciences: The Critical Background to Modern Science, 1800–1905.* Colin Howson (ed.) (Cambridge: Cambridge University Press), pp. 309–39 (see also 1976, ed.).

(1976h) 'Consolations for the Specialist' (in Italian), *Critica e crescita della conoscenza.* Imre Lakatos and Alan Musgrave (eds.). A cura di Giulio Giorello (Milano: Feltrinelli). Reprinted 1986.

(1976i) 'Cómo ser un buen empirista. Defensa de la tolerancia en cuestiones epistemológicas', translated by Diego Ribes and Maria Rosario de Madaria. With an introduction by Diego Ribes (Col. Cuadernos teorema, 7) (Valencía, Fac. Filosofia y Letras), 62 pp. (see also 1963d, 1968e, 1970h, 1970l, 1972c, 1979c, 1980i, 1982g).

(1976j) *Filosofia de la ciencia y religion.* Sigueme: Salamanca, 156 pp.

(1977a) 'Marxist Fairytales From Australia', *Inquiry* (20) 2–3, pp. 372–97 (see also 1978B, 154–82).

(1977b) 'Changing Patterns of Reconstruction: A Review of W. Stegmüller's *Theorienstrukturen und Theoriendynamik*', *British Journal for the Philosophy of Science* (28) 4, pp. 351–82.

(1977c) 'Unterwegs zu einer dadaistischen Erkenntnistheorie', *Unter dem Pflaster liegt der Strand* (4), Hans Peter Duerr (ed.) (Berlin: Karin Kramer Verlag), pp. 9–88.

(1977d) 'Rationalism, Relativism and Scientific Method', *Philosophy in Context*, Supplement (6) 1, pp. 7–19 (see also 1978B, 210–17, and this volume).

(1977e) 'Philosophy Today', *Teaching Philosophy Today*. Terrell Ward Bynum and Sidney Reisberg (eds.) (Ohio: Bowling Green State University), pp. 72–5.

(1977f) Willy Hochkeppel im Gespräch mit Paul Feyerabend am 19.12.1977. Ms.

(1978a) 'Life at the LSE?', *Erkenntnis* (13) 2, pp. 297–304 The author's name of this piece appears as 'Fantomas'. (See also 1978A, 210–17).

(1978b) 'From Incompetent Professionalism to Professionalized Incompetence – The Rise of a New Breed of Intellectuals', *Philosophy of the Social Sciences* (8) 1, pp. 37–53 (see also 1978B, 183–209).

(1978c) 'Reply to Tibbetts and Hattiangadi', *Ibid.* (8) 2, pp. 184–6.

(1978d) 'Kleines Gespräch über große Worte', *Unter dem Pflaster liegt der Strand* (5). Hans Peter Duerr (ed.) (Berlin: Karin Kramer Verlag), pp. 123–39 (see also 1978B, pp. 255–72).

(1978e) 'In Defense of Aristotle. Comments on the Condition of Content Increase', *Progress and Rationality in Science, Boston Studies in the Philosophy of Science, Volume 58*, Gerard Radnitzky and Gunnar Andersson (eds.) (Dordrecht and Boston: D. Reidel), pp. 143–80 (see also 1980d).

(1978f) 'The Gong Show – Popperian Style', *Ibid.*, pp. 387–92.

(1978g) 'Das Märchen Wissenschaft: Plädoyer für einen Supermarkt der Ideen', *Kursbuch, Utopien II. Lust an der Zukunft, Heft 53*. Karl Markus Michel and Harald Wieser (eds.) (Berlin: Kursbuch Verlag), pp. 47–70.

(1978h) 'Jedermann sollte die Möglichkeit haben, seine Wahrheit, seine Falschheit für sich selbst zu wählen', *Tagesanzeiger*: Zürich, 15.7.1978.

(1979a) 'Reply to Hellman's Review', *Metaphilosophy* (10) 2, pp. 202–6.

(1979b) 'To the Editors: Science and Society: An Exchange' by Daniel J. Kevles, Paul K. Feyerabend and David Joravsky, *New York Review of Books*, 11 October pp. 48–50.

(1979c) *Jak Byc Dubrym Empirysta*. Krystna Zamiare (ed.) Panstuome Wydawnict wo Naukoae: Warszana. 252 pp. (see also 1963d, 1968e, 1970h, 1970l, 1972c, 1976i, 1980i, 1982g).

(1979d) 'Dialogue on Method', *The Structure and Development of Science*. Gerard Radnitzky and Gunnar Andersson (eds.) (Dordrecht and Boston: D. Reidel), pp. 63–131 (see also 1989A, 1976e, 1981f).

(1979e) 'El mito de la "ciencia" y su papel en la sociedad' translated by Angel Barahona, *Cuadernos teorema*. Departamento de Logica de la Universidad de Valencia, Spain, pp. 11–28.

(1979f) 'Vorwort von Paul Feyerabend', *Die Selbstorganisation des Universums – Vom Urknall zum menschlichen Geist*. Erich Jantsch (München: Karl Hanser Verlag). Reprinted 1982 (München: Deutscher Taschenbuchverlag), pp. 13–15.

(1980a) 'Democracy, Elitism, and Scientific Method', *Inquiry* (23) 1, pp. 3–18 (reprinted in this volume).

(1980b) 'Zahar on Mach, Einstein and Modern Science', *British Journal for the Philosophy of Science* (31) 3, pp. 273–82 (see also 1981B, 89–98).

(1980c) 'A lógica, o bê-a-bá e o Professor Gellner', translated by Balthazar Barbosa Filho, *Cuadernos de história e filosofia da ciência*. Campinas (n.1) pp. 77–89.

(1980d) 'Eine Lanze für Aristoteles: Bemerkungen zum Postulat der Gehaltvermeh-rung', *Fortschritt und Rationalität der Wissenschaft. Die Einheit der Gesellschaftswis-*

senschaften, Band 24. Gerard Radnitzky and Gunnar Andersson (eds.) (Tübingen: J. C. B. Mohr (Paul Siebeck)), pp. 157–98 (see also 1978e).

(1980e) 'Watkins' Kommentar – ein Musterbeispiel rationalistischer Kritik', *Ibid.*, pp. 441–7.

(1980f) 'How to Defend Society Against Science', *Introductory Readings in the Philosophy of Science.* E. D. Klemke, Robert Hollinger and David A. Kline (eds.) (Buffalo, N.Y.: Prometheus Books), pp. 55–65 (see also 1975b, 1981c, 1984c, 1988g).

(1980g) 'Warum Platon? Ein kleines Gespräch', *ETH-Bulletin*, no.158, October (see also 1981a, 1993m).

(1980h) 'Galileo's Observations' (Letter to the editors), *Science.* Volume 209, p. 544.

(1980i) 'Comment être un bon empiriste. Plaidoyer en faveur de la tolérance en matière épistemologique', *De Vienne à Cambridge: L'héritage du positivisme logique de 1950 à nos jours.* Pierre Jacob (ed.) (Paris: Éditions Gallimard), pp. 245–76 (see also 1963d, 1968e, 1970h, 1970l, 1972c, 1976i, 1979c, 1982g).

(1981a) 'Warum Platon? Ein kleines Gespräch', *Unter dem Pflaster liegt der Strand* (8) Hans Peter Duerr (ed.) (Berlin: Karin Kramer Verlag), pp. 177–87 (see also 1980g, 1993m).

(1981b) 'More Clothes from the Emperor's Bargain Basement. A review of *Progress and its Problems* by Larry Laudan', *British Journal for the Philosophy of Science* (32) 1, pp. 57–71 (see also 1981B, 231–46).

(1981c) 'How to Defend Society Against Science', *Scientific Revolutions.* Ian Hacking (ed.) (Oxford: Oxford University Press), pp. 156–67. Reprinted 1983, 1985 (see also 1975b, 1980f, 1984c, 1988g).

(1981d) 'Irrationalität oder: Wer hat Angst vorm schwarzen Mann?', *Der Wissenschaftler und das Irrationale, Beiträge aus Philosophie und Psychologie, Band 2.* Hans Peter Duerr (ed.) (Frankfurt am Main: Syndikat), pp. 37–59 (see also 1985e, 1985f).

(1981e) 'Foreword', *Radical Knowledge. A Philosophical Inquiry into the Nature and Limits of Science.* Gonzalo Munévar (ed.) (Indianapolis, IN: Hackett), pp. ix–x.

(1981f) 'Über die Methode. Ein Dialog', *Voraussetzungen und Grenzen der Wissenschaft. Die Einheit der Gesellschaftswissenschaften, Band 25.* Gerard Radnitzky and Gunnar Andersson (eds.) (Tübingen: J. C. B. Mohr (Paul Siebeck)), pp. 175–253 (see also 1989A, 1976e, 1979d).

(1981g) 'Rückblick', *Versuchungen. Aufsätze zur Philosophie Paul Feyerabends, Band II.* Hans Peter Duerr (ed.) (Frankfurt am Main: Suhrkamp), pp. 320–72.

(1981h) 'Mentale Ereignisse und das Gehirn', *Analytische Philosophie des Geistes. Philosophie Analyse und Grundlegung, Band 6.* Peter Bieri (ed.) (Meisenheim and Königstein: Verlag Anton Hain), pp. 121–2 (see also 1963f, 1970h, 1990d, 1993n).

(1981i) 'Foreword', *Chemistry, Quantum Mechanics and Reductionism. Perspectives in Theoretical Chemistry. Lecture Notes in Chemistry. Volume 24.* Hans Primas (Berlin, Heidelberg and New York: Springer-Verlag), pp. v–viii (see also 1983g).

(1981j) 'Galileo as a Scientist' (Letter to the editors), *Science.* Volume 211, 27 February.

(1981k) 'Bemerkungen zur Angst vor der Irrationalität', *Der Wissenschaftler und das Irrationale.* Vol. 1, Hans Peter Duerr (ed.) (Frankfurt am Main: Syndikat).

(1981l) Interview by Teresa Ordunya which appeared on 4.17.98 at feyerabend @lists.village.virginia.edu

(1982a) 'Die Aufklärung hat noch nicht begonnen', *Von der Verantwortung des Wissens:*

ERIC OBERHEIM

Positionen der neueren Philosophie der Wissenschaft. Paul Good (ed.) (Frankfurt am Main: Suhrkamp), pp. 24–40.

(1982b) 'Lo spettro del relativismo' translated by Eugen Galasso, *Lo scienziato e il filosofo* (Milano: Laboratorio di ricerche anarchiche), 4/87, pp. 87–95.

(1982c) 'Academic Ratiofascism. Comments on Tibor Machan's Review', *Philosophy of the Social Sciences* (12) 2, pp. 191–5.

(1982d) 'Redet nicht herum – organisiert Euch', *Unter dem Pflaster liegt der Strand* (10) Hans Peter Duerr (ed.) (Berlin: Karin Kramer Verlag), pp. 169–74.

(1982e) 'The Strange Case of Astrology', *Philosophy of Science and the Occult.* Patrick Grim (ed.) (Albany, NY: State University of New York Press), pp. 19–24. Reprinted 1990. (See also 1978B, Pt. 2, ch. 6.)

(1982f) 'Science – Political Party or Instrument of Research?', *Speculations in Science and Technology* (5) 4, pp. 343–52.

(1982g) *Come essere un buon empirista.* Roma, Borla, 136 pp. (see also 1963d, 1968e, 1970h, 1970l, 1972c, 1976i, 1979c, 1980i).

(1983a) 'Feyerabend vs. Gardner: Science: Church or Instrument of Research?', *Free Inquiry* (3) pp. 58–60.

(1983b) 'Wissenschaft als Kunst', *Psychologie Heute* (10) 9, pp. 56–62.

(1983c) 'Some Observations on Aristotle's Theory of Mathematics and of the Continuum', *Contemporary Perspectives on the History of Philosophy. Midwest Studies in Philosophy, Volume 8.* Peter A. French, Theodore E. Uehling, Jr and Howard K. Wettstein (eds.) (Minneapolis: University of Minnesota Press), pp. 67–88 (see also 1987A, 219–46).

(1983d) 'Eindrücke 1945–1954', *Vom Reich zu Österreich.* J. Jung (ed.) (Salzburg and Wien: Residenz Verlag), pp. 263–71 (see also 1984h).

(1983e) 'Die Wahrheit, das laute Kind', Paul Feyerabend on Albrecht Fölsing: 'Galileo Galilei – Prozeß ohne Ende. Eine Biographie', *Der Spiegel*, 37. Jg. Nr. 28, 11th July 1983, pp. 141–3.

(1983f) 'Noch einmal: Können Hexen fliegen?' by Stanislaw Lem and Paul Feyerabend, *Der gläserne Zaun. Aufsätze zu Hans Peter Duerrs >>Traumzeit<<.* Rolf Gehlen and Bernd Wolf (eds.) (Frankfurt am Main: Syndikat), pp. 232–42.

(1983g) 'Foreword', *Chemistry, Quantum Mechanics and Reductionism. Perspectives in Theoretical Chemistry.* Hans Primas (Berlin, Heidelberg, New York and Tokyo: Springer Verlag), pp. v–viii. 2nd revised edition (see also 1981i).

(1984a) 'Xenophanes: A Forerunner of Critical Rationalism?' translated by Dr. John Krois, *Rationality in Science and Politics. Boston Studies in the Philosophy of Science, Volume 79.* Gunnar Andersson (ed.) (Dordrecht and Boston: D. Reidel), pp. 95–109.

(1984b) 'Philosophy of Science 2001', *Methodology, Metaphysics and the History of Science. In memory of Benjamin Nelson. Boston Studies in the Philosophy of Science, Volume 84.* Robert S. Cohen and Marx W. Wartofsky (eds.) (Dordrecht and Boston: D. Reidel), pp. 137–47.

(1984c) 'Come difendere la società contro la scienza' translated by Libero Sosio, *Rivoluzioni scientifiche.* Ian Hacking (ed.) (Roma and Bari: Laterza e Figli), pp. 209–88 (see also 1975b, 1980f., 1981c, 1984c, 1988g).

(1984d) 'Mach's Theory of Research and its Relation to Einstein', *Studies in History and Philosophy of Science* (15) 1, pp. 1–22 (see also 1987A, 192–218, 1988a).

(1984e) 'Was heißt das, wissenschaftlich zu sein?', *Zeitschrift für Parapsychologie und Grenzgebiete der Psychologie* (26) pp. 58–64 (see also 1985, ed.).

(1984f) 'On the Limits of Research', *New Ideas in Psychology. An International Journal of Innovative Theory in Psychology* (2) 1, pp. 3–7.

(1984g) 'The Lessing Effect in the Philosophy of Science: Comments on Some of my Critics', *Ibid.* (2) 2, pp. 127–36.

(1984h) 'Eindrücke 1945–1954', *Unter dem Pflaster liegt der Strand* (13), Hans Peter Duerr (ed.) (Berlin: Karin Kramer Verlag), pp. 81–7 (see also 1983d).

(1984i) 'Interview mit Paul Feyerabend', *Ibid.*, pp. 89–99.

(1984j) 'Interview', *Entretiens avec Le Monde 3. Idées Contemporaines.* Éditions la Découverte et *Journal Le Monde*, Paris (Guitta Pessis Pasternak, 28 February 1982), pp. 27–34.

(1984k) 'Betrayers of the Truth: Fraud and Deceit in the Halls of Science', *New York Review of Books* (30) 19 January pp. 21–2.

(1984l) *Adios a la razon.* Cuadernos de filosofia y ensayo (Madrid: Tecnos), 195 pp.

(1984m) 'Science as Art', *Art and Text.* nos. 12 and 13 (Australia: University of Melbourne), pp. 16–46.

(1985a) 'Kleine Anleitung zum Denken – speziell geschrieben für Herrn Stanislaw Lem', *Unter dem Pflaster liegt der Strand* (14), Hans Peter Duerr (ed.) (Berlin: Karin Kramer Verlag), pp. 133–5.

(1985b) 'Galileo and the Tyranny of Truth', *The Galileo Affair: A Meeting of Faith and Science: Proceedings of the Cracow Conference 24 to 27 May 1984.* G. V. Cogne, M. Hellor and J. Zycinski (eds.) (Città del Vaticano: Specola Vaticana), pp. 155–66 (see also 1987A, 247–64).

(1985c) 'Reply', *Ibid.*, pp. 165–6.

(1985d) 'Autoritäre Konstruktionen oder demokratische Entschlüsse? Bemerkungen zum Problem des kulturellen Pluralismus', *Objektivationen des Geistigen: Beiträge zur Kulturphilosophie in Gedenken an Walther Schmied-Kowarzik (1885–1954).* Wolf-dietrich Schmied-Kowarzik (ed.) (Berlin: Dietrich Reimer Verlag), pp. 137–44.

(1985e) 'Irrationalität oder: Wer hat Angst vorm schwarzen Mann?', *Der Wissenschaftler und das Irrationale, Band 3, Beiträge aus der Philosophie.* Hans Peter Duerr (ed.) (Frankfurt am Main: Syndikat), pp. 33–55 (see also 1981d, 1985f).

(1985f) 'L'irrazionalità ovvero: chi ha paura dell'uomo nero?' translated by Francesca Beltrami. *Aut Aut* (n. 205) pp. 67–86 (see also 1981d, 1985e).

(1985g) 'Feyerabend racconta Feyerabend', *Scienza Esperienza*, Marzo, pp. 15–22.

(1986a) 'Eingebildete Vernunft: Die Kritik des Xenophanes an den Homerischen Göttern', *Zur Kritik der wissenschaftlichen Rationalität. Zum 65. Geburtstag von Kurt Hübner.* Hans Lenk (ed.) (Freiburg and München: Verlag Karl Alber), pp. 205–23.

(1986b) 'Progress and Reality in the Arts and in the Sciences', *Progress in Science and Its Social Conditions. Proceedings of a Nobel Symposium 58, Held at Lidingoe, Sweden, 15–19 August 1983.* Tord Ganelius (ed.) (Oxford: Pergamon Press), pp. 223–33 (see also 1987A, 143–61).

(1986c) 'Science and Ideology: A Response to Rollin', *New Ideas in Psychology* (4) 2, pp. 153–8.

(1986d) 'Trivializing Knowledge: A Review of Popper's *Postscript*', *Inquiry* (29) 1, pp. 93–119 (see also 1987A, 162–91).

(1986e) 'Philosophie in einer wissenschaftsgetränkten Kultur', *Enzyklopädie der Erziehungswissenschaften*. Marburg, pp. 157–62.

(1987a) 'Creativity – A Dangerous Myth', *Critical Inquiry* (13) 4, pp. 700–11 (see also 1987A, 128–42).

(1987b) 'Putnam on Incommensurability', *British Journal for the Philosophy of Science* (38) 1, pp. 75–81 (see also 1987A, 265–72).

(1987c) 'Cultural Pluralism or Brave New Monotony', *The Culture of Fragments: The Journal of The Columbia University Graduate School of Architecture Planning and Preservation*. Gianmarco Vergani (ed.) Precis (6) (New York: Rizzoli International), pp. 35–45 (see also 1987A, 273–9).

(1987d) 'Reason, Xenophanes and the Homeric Gods', *The Kenyon Review. New Series* (9) 4, pp. 12–22 (see also 1987A, 90–102).

(1987e) 'Galileo e la tirannia della verità', *Astronomia: mensile di scienza e cultura* (17) Novembre (Verona: Konus Italia), pp. 28–36.

(1988a) 'Machs Theorie der Forschung und ihre Beziehung zu Einstein', *Ernst Mach. Werk und Wirkung*. Rudolf Haller and Friedrich Stadler (eds.) (Wien: Verlag Hölder-Pichler-Tempsky), pp. 435–62 (see also 1987A, 192–218, 1984d).

(1988b) 'Knowledge and the Role of Theories', *Philosophy of the Social Sciences* (18) 2, pp. 157–78 (see also 1987A, 103–27, 1995d).

(1988c) 'Perché essere 'scientifico'?', *La svolta relativistica nell'epistemologia contemporanea*. Rosaria Egidi (ed.) (Milano: Franco Angeli Libri), pp. 87–93.

(1988d) 'Creativity', *Speciale sabato*, Giò Rezzonico (ed.) Eco di Locarno sabato 1 Ottobre 1988.

(1988e) 'Galileo e gli dèi omerici' translated by Gianni Rigamonti, *Nuove effemeridi* (2) pp. 54–9.

(1988f) 'Reply to Agassi', *The Gentle Art of Philosophical Polemics: Selected Reviews and Comments*. Joseph Agassi (ed.) (La Salle, IL: Open Court), pp. 405–16 (see also 1978B, 1976b).

(1988g) 'How to Defend Society Against Science', *Introductory Readings in the Philosophy of Science*. Revised Edition. E. D. Klemke, Robert Hollinger and David A. Kline (eds.) (Buffalo, NY: Prometheus Books), pp. 34–44 (see also 1975b, 1980f, 1981c, 1984c).

(1988h) 'Abstrakte oder persönliche Rechtfertigung', *Streitbare Philosophie. Margherita von Brentano zum 65. Geburtstag*. Gabriele Althaus and Irmingard Staeuble (eds.). Berlin, pp. 53–67.

(1989a) 'Realism and the Historicity of Knowledge', *The Journal of Philosophy* (86) 8, pp. 393–406 (see also 1990a, and *The Conquest of Abundance*).

(1989b) 'Antilogikē', *Freedom and Rationality: Essays in Honor of John Watkins From his Colleagues and Friends. Boston Studies in the Philosophy of Science, Volume 117*. Fred D'Agostino and I. C. Jarvie (eds.) (Dordrecht, Boston and London: Kluwer Academic Publishers), pp. 185–9.

(1989c) 'Wem nützt die Wissenschaftstheorie?', *Traditionen und Perspektiven der analytischen Philosophie. Festschrift für Rudolf Haller*. Wolfgang L. Gombocz, Heiner Rutte and Werner Sauer (eds.) (Wien: Hölder-Pichler-Tempsky), pp. 568–74.

(1989d) *Limites de la ciencia: explicacion, reduccion y empirismo* (Barcelona: Ediciones Paidos).

(1989e) 'Erkenntnistheorie, anarchische', *Handlexikon zur Wissenschaftstheorie*. Helmut Seiffert and Gerard Radnitzky (eds.) (München: Ehrenwirth), pp. 58–61.

(1989f) 'Rationalismus', *Ibid.*, pp. 280–2.

(1989g) 'Relativismus', *Ibid.*, pp. 292–6.

(1990a) 'Realism and the Historicity of Knowledge', *Third International Locarno Conference on Science and Society, October 1988* (Canton, MA: Science History Publications), pp. 142–53 (see also 1989a).

(1990b) 'Grandi intuizioni e non piccoli esperimenti', *Riza scienze: La banca delle idee: Pittori, scienziati, psicoterapeuti, giornalisti, scrittori, si confrontano sulla creatività e i suoi segreti (39) Mensili di scienza dell'uomo.* A cura di G. Borrini, pp. 58–67.

(1990c) 'Il realizmo e la storicità della conoscenza', translated by Pierluigi Barrotta, *Nuova civiltà delle macchine*, Roma (8) 2–3, pp. 21–8.

(1990d) 'Mental Events and the Brain', *Mind and Cognition: A Reader.* William G. Lycan (ed.) (Oxford: Blackwell), pp. 295–6 (see also 1963f, 1970h, 1981h, 1993n).

(1991a) 'Concluding Unphilosophical Conversation', *Beyond Reason: Essays on the Philosophy of Paul Feyerabend. Boston Studies in the Philosophy of Science, Volume 132.* Gonzalo Munévar (ed.) (Dordrecht, Boston and London: Kluwer Academic Publishers), pp. 487–527.

(1991b) 'Gods and Atoms: Comments on the Problem of Reality', *Thinking Clearly about Psychology Volume 1. Matters of Public Interest. Essays in honor of Paul E. Meehl.* Dante Cicchetti and William M. Grove (eds.) (Minneapolis: University of Minnesota Press), pp. 91–9.

(1991c) 'Dialogo con la natura', translated by Maria Paola Arena, *Prometeo: Rivista trimestrale di scienze e storia.* Anno 9, Numero 33, Marzo, pp. 6–13.

(1991d) 'It's Not Easy to Exorcize Ghosts: Commentary on Mario Bunge', *New Ideas in Psychology* (9) 2, pp. 181–6.

(1991e) 'Etica como medida de la verdad científia', *Feyerabend y algunas metodologías de la investigacíon.* Ana María Tomeo (ed.) (Milán, Montevideo, Uruguay: Nordan-Comunidad), pp. 229–42 (see also 1991f, 1992a, and *The Conquest of Abundance*).

(1991f) 'L'etica come misura di verità scientifica', translated by Davide Sparti, *Iride: Filosofia e discussione pubblica* (7) pp. 68–76 (see also 1991e, 1992a, and *The Conquest of Abundance*).

(1991g) 'Monster in the Mind', *The Guardian.* Friday, 13 December 1991, p. 26.

(1992a) 'Ethics as a Measure of Scientific Truth', *From the Twilight of Probability: Ethics and Politics.* William R. Shea and Antonio Spadafora (eds.) (Canton, MA: Science History Publications), pp. 106–14 (see also 1991e, 1991f, and *The Conquest of Abundance*).

(1992b) 'Historical Comments on Realism', *International Conference on Bell's Theorem and the Foundations of Modern Physics.* A. van der Merve, F. Selleri and G. Tarozzi (eds.) (Singapore: World Scientific Pub.), pp. 194–202.

(1992c) 'Il relativismo ontologico', translated by Roberta Lanfredini, *Iride: Filosofia e discussione pubblica* (8) pp. 7–20.

(1992d) Review of *Science and Relativism: Some Key Controversies in the Philosophy of Science* by Larry Laudan, *Isis* (83) 2, pp. 367–8.

(1992e) 'Quantum Theory and our View of the World', *Stroom: Mededelingenblad Faculteit Natuur- and Sterrenkunde* (6) 28 (Universiteit Amsterdam), pp. 19–24 (see also 1994q, and *The Conquest of Abundance*).

(1992f) 'Call for Papers: ix, x, xi', *Common Knowledge* (1) 1, pp. 8–10.

(1992g) 'Atoms and Consciousness', *Ibid.*, pp. 28–32.

(1992h) Little Review of *Conversations on the Dark Secrets of Physics* by Edward Teller, *Ibid.*, p. 152.

(1992i) Little Review of *Models of My Life* by Herbert A. Simon, *Ibid.*, p. 153.

(1992j) 'Response to Toulmin', *Ibid* (1) 2, p. 16.

(1992k) Little Review of *Galileo's Revenge* by Peter W. Huber, *Ibid.*, p. 130.

(1992l) 'Nature as a Work of Art: A Fictitious Lecture Delivered to a Conference Trying to Establish the Increasing Importance of Aesthetics for Our Age', *Ibid.*, (1) 3, pp. 3–9 (see also 1994o).

(1992m) Little Review of *Origins: The Lives and Worlds of Modern Cosmologists* by Alan Lightman and Roberta Brauer, *Ibid.*, pp. 161–2.

(1992n) Little Review of *History of American Cinema, Ibid.*, p. 162.

(1992o) 'Erkenntnis ohne Theorie – Vom Nutzen der Abstraktion und vom Recht des Besonderen', *Lettre International. Volume 16.* pp. 66–71.

(1993a) Little Review of *The Edges of the Earth in Ancient Thought: Geography, Exploration, and Fiction* by James Romm, *Common Knowledge* (2) 1, p. 125.

(1993b) Little Review of *Particles and Waves: Historical Essays in the Philosophy of Science* by Peter Achinstein, *Ibid.*, pp. 126–7.

(1993c) Little Review of *The Laser in America, 1950–70* by Joan Lisa, *Niels Bohr's Times in Physics, Philosophy and Polity* by Abraham Pais, and *Science as Practice and Culture* by Andrew Pickering, *Ibid.*, p. 127.

(1993d) 'Intellectuals and the Facts of Life', *Ibid.*, (2) 3, pp. 6–9 (see also *The Conquest of Abundance*).

(1993e) Little Review of *The Beginnings of Western Science: The European Scientific Tradition in Philosophical, Religious, and Institutional Context, 600 B.C. to A.D. 1450* by David C. Lindberg, *Ibid.*, p. 141.

(1993f) Little Review of *Travel Writing and Transculturation* by Mary Louise Pratt, *Ibid.*, pp. 141–2.

(1993g) Little Review of *Engineering and the Mind's Eye* by Eugene S. Ferguson, *Ibid.*, p. 142.

(1993h) 'L'arte prodotto della natura come opera d'arte', *Notiziario dell'enea. Energia e innovazione.* Fausto Borrelli (ed.) 2–3 Febbraio-Marzo, pp. 53–64 (see also 1994n, 1995c, and *The Conquest of Abundance*).

(1993i) 'Die Natur als ein Kunstwerk', *Die Aktualität des Ästhetischen*, Wolfgang Welsch (ed.) (München: Wilhelm Fink Verlag), pp. 278–87.

(1993j) 'The End of Epistemology?', *Physics, Philosophy and Psychoanalysis: Essays in Honor of Adolf Grünbaum. Boston Studies in the Philosophy of Science, Volume 76.* R. S. Cohen and Larry Laudan (eds.) (Dordrecht and Boston: D. Reidel), pp. 187–204.

(1993k) 'Sobre la diversidad de la ciencia', *Imágenes y metáforas de la ciencia*. Lorena Preta (ed.) (Madrid: Alianza Editorial), pp. 152–9.

(1993l) 'Materialism and the Mind–Body Problem', *Folk Psychology and the Philosophy of Mind.* Scott M. Christensen and Dale R. Turner (eds.) (Hillsdale, NJ: Lawrence Erlbaum Associates), pp. 3–16 (see also 1981A, 161–75, 1963g, 1967k, 1969e, 1970i).

(1993m) *Por que no Platon?* Col. Cuadernos de filosofia y ensayos (Madrid: Tecnos), 188 pp. (see also 1980g, 1981a).

(1993n) 'Mentale Ereignisse und das Gehirn', *Analytische Philosophie des Geistes. 2, verbesserte Auflage. Neue Wissenschaftliche Bibliothek.* Peter Bieri (ed.) (Bodenheim: Athenäum Hain Hanstein), pp. 121–2 (see also 1963f, 1970h, 1981h, 1990d).

(1993o) 'Ein absolutes Prinzip Vielfalt wäre Idiotie – Interview mit Gregor Schiemann', *Freitag.* 19 Feb. Nr. 8, 6.

(1993) 'Not a Philosopher', *Falling in Love with Wisdom: American Philosophers Talk about their Calling.* D. D. Karnos and R. G. Shoemaker (eds.) (New York: Oxford University Press), pp. 16–7.

(1994a) 'Realism', *Artifacts, Representations and Social Practice. Essays for Marx Wartofsky. Boston Studies in the Philosophy of Science, Volume 154.* Carol C. Gould and Robert S. Cohen (eds.) (Dordrecht, Boston and London: Kluwer), pp. 205–22.

(1994b) 'Idee: balocchi intellettuali o guide per la vita?', translated by Antonella Iocca, *Il caso e la libertà.* Mauro Cerruti, Paolo Fabbri, Giulio Giorello and Lorena Preta (eds.) (Roma and Bari: Laterza e Figli), pp. 28–53.

(1994c) 'Contro l'ineffabilità culturale', translated by Guido Lagomarsino, *Volontà: Tutto è relativo, o no?*, no. 2/3 anno 48. Volontà: Laboratorio di Ricerche Anarchiche (ed.) (Milano: Edizioni Volontà), pp. 97–106.

(1994d) Little Review of *La Mettrie: Medicine, Philosophy and Enlightenment* by Kathleen Wellman, *Common Knowledge* (3) 1, p. 147.

(1994e) Little Review of *The Janus of Genius* by B. J. T. Dobbs, *Ibid.*, p. 149.

(1994f) Little Review of *Out of The Shadows: Herschel, Talbot and the Invention of Photography* by Larry J. Schaaf, *Ibid.*

(1994g) Little Review of *Printing the Written Word: The Social History of Books, Circa 1450–1520* by Sandra C. Hindman, *Ibid.*, pp. 149–50.

(1994h) Little Review of *Conceptual Revolutions* by Paul Thagard, and *Simulating Science: Heuristics, Mental Models,* and *TechnoScientific Thinking* by Michael E. Gorman, *Ibid.*, p. 150.

(1994i) Little Review of *Wild Knowledge: Science, Language and Social Life in a Fragile Environment* by Will Wright, *Ibid.*, p. 151.

(1994j) 'Potentially Every Culture is All Cultures', *Ibid* (3) 2, pp. 16–22.

(1994k) Little Review of *Novelties in the Heavens: Rhetoric and Science in the Copernican Controversy* by Jean Dietz Moss, *Ibid.*, p. 158.

(1994l) 'Concerning an Appeal for Philosophy', *Ibid.*, (3) 3, pp. 10–13 (see also *The Conquest of Abundance*).

(1994m) Little Review of *Galileo Courtier: The Practice of Science in the Culture of Absolutism* by Mario Biagioli, and *Reconstructing Scientific Revolutions* by Paul Hoyningen-Huene, *Ibid.*, p. 173.

(1994n) 'Art as a Product of Nature as a Work of Art', *World Futures* (40) 1–3, pp. 87–100 (see also 1993h, 1995c, and *The Conquest of Abundance*).

(1994o) 'Natur als Werk der Kunst – Fiktiver Vortrag über die wachsende Bedeutung der Ästhetik', *Lettre International. Volume 25*, pp. 40–2 (see also 1992l).

(1994p) 'Has the Scientific View of the World a Special Status, Compared with Other Views?', *Physics and Our View of the World,* J. Hilgevoord (ed.) (Cambridge: Cambridge University Press), pp. 135–48.

(1994q) 'Quantum Theory and Our View of the World', *Ibid.*, pp. 149–68 (see also 1992e, and *The Conquest of Abundance*).

(1995a) 'Università e primi viaggi: Un' autobiografia', *Atque: materiali tra filosofia e psicoterapia.* P. F. Pieri (ed.) (Bergamo: Moretti and Vitali), pp. 9–26.

(1995b) 'Quale realità?', translated by Alessandro Pagnini, *Storia della filosofia, storia della scienza: Saggi in onore di Paolo Rossi.* Antonello La Vergata und Alessandro Pagnini (eds.) (Scandicci and Firenze: La Nuova Italia Editrice), pp. 79–91.

(1995c) 'Art as a Product of Nature as a Work of Art', *Science, Mind and Art: Essays on Science and the Humanistic Understanding in Art, Epistemology, Religion, and Ethics. In Honor of Robert S. Cohen. Boston Studies in the Philosophy of Science, Volume 165.* K. Gavroglu, J. Stachel and M. W. Wartofsky (eds.) (Dordrecht, Boston and London: Kluwer Academic Publishers), pp. 1–18 (see also 1993h, 1994n, and *The Conquest of Abundance*).

(1995d) 'Knowledge and the Role of Theories', *Pragmatik: Handbuch Pragmatischen Denkens, Vol 5. Pragmatische Tendenzen in der Wissenschaftstheorie.* Herbert Stachowiak (ed.) (Hamburg: Felix Meiner Verlag), pp. 59–80 (see also 1988b).

(1995e) 'Il relativismo ontologico', translated by Roberta Lanfredini, *Realismo/Antirealizmo: Aspetti del dibattito epistemologico contemporaneo,* Alessandro Pagnini (ed.) (Scandicci and Firenze: La Nuova Italia Editrice), pp. 39–57.

(1995f) 'Two Letters of Paul Feyerabend to Thomas S. Kuhn on a Draft of *The Structure of Scientific Revolutions*', Paul Hoyningen-Huene (ed.) *Studies in History and Philosophy of Science* (26) 3, pp. 353–87.

(1995g) 'Bohr, Niels', *The Oxford Companion to Philosophy.* Ted Honderich (ed.) (Oxford and New York: Oxford University Press), p. 98.

(1995h) 'Mach, Ernst', *Ibid.*, p. 516.

(1995i) 'Philosophy: World and Underworld', *Ibid..*, p. 678.

(1995j) 'Science, History of the Philosophy of', *Ibid.*, pp. 806–9.

(1995k) 'Universals as Tyrants and as Mediators', *Critical Rationalism, Metaphysics and Science. Essays for Joseph Agassi, Volume I. Boston Studies in the Philosophy of Science, Volume 161.* I. C. Jarvie and N. Laor (eds.) (Dordrecht, Boston and London: Kluwer Academic Publishers), pp. 3–14 (see also *The Conquest of Abundance*).

(1995l) 'Perché non separare la scienza dello stato?', *Paul Feyerabend nella stampa periodica italiana 1975–95.* Luciana Dini Vecchio (ed.) Centro di Documentazione e Ricerca. Instituto Italiano per GCI Studi Filosofici, pp. 9–10.

(1995m) 'Conoscere per sopravvivere: scienziati, tecnologie, esperti dominano la nostra vita quotidiana, ma chi controlla la scienza? E' possibile liberarsi da questo fascino e potere?', *Ibid.*, pp. 73–80.

(1995n) 'Mill e Hegel: due modelli di cambiamento epistemico', *Ibid.*, pp. 86–92.

(1995o) 'Meglio imparar l'arte', *Ibid.*, p. 105.

(1995p) 'Anteprima. Ma la chiesa fu più razionale', *Ibid.*, p. 114.

(1995q) 'Dialogo con la natura: Diverse concezioni del mondo sono ugualmente reali nella prospettiva di un relativismo ontologico', *Ibid.*, pp. 137–40.

(1995r) 'L'etica come misura di verità scientifica', *Ibid.*, pp. 144–50.

(1995s) 'Miseria dell'epistemologia', *Ibid.*, pp. 154–61.

(1995t) 'Conto l'ineffabilità cultrale. Oggettivismo, relativismo e altre chimere', *Ibid.*, pp. 211–15.

(1995u) 'Separare la scienza dallo stato', *Ibid.*, p. 198.

(1995v) 'Ma che razza di idee sono se non aiutano a vivere?', *Ibid.*, pp. 205–6.

(1995w) *Imre Lakatos, Paul K. Feyerabend, sull'orlo della scienza. Pro e contro il metodo.* Matteo Motterlini (ed.) (Milano: Raffaello Cortina Editore).

(1995x) *Paul Feyerabend. Briefe an einen Freund.* Hans Peter Duerr (ed.) (Frankfurt am Main: Suhrkamp), 291 pp.

(1995y) 'Three Interviews with Paul Feyerabend', conducted by Vittorio Hoesle, *Telos. A Quarterly Journal of Critical Thought* (102) pp. 115–49.

(1995z) 'Erkenntnis und Praxis – Gefahren der Abstraktion, Fallgruben des Realen', *Lettre International.* Sommer, pp. 20–3. (The text is based on an interview with an Italian television station.)

(1995aa) 'Zum Leben braucht man die Nähe zu den Menschen – Paul Feyerabend im Gespräch mit Matthias Kroß', *Information Philosophie. Heft 1.*, pp. 28–32. (First published in the Berlin news magazine *Zitty.*)

(1996a) 'Contro l'ineffabilità culturale. Oggettivismo, relativismo e altre chimere' translated by Maria Grazia Ciani, *Universalità and differenza. Cosmopolitismo e relativismo nelle relazioni tra identità sociali e culture* (Milano: FrancoAngeli), pp. 27–35.

(1996b) 'Ambiguità e armonia', translated by Caterina Castellani, *Lezioni trentine. A cura di Francesca Castellani* (Roma and Bari: Laterza e Figli).

(1996c) 'Theoreticians, Artists and Artisans', *Leonardo: Journal of the International Society for the Arts, Sciences and Technology* (29) 1, pp. 23–8.

(1996d) 'Il calderone delle streghe', *L'almanacco dell'Altana* (Roma: Edizioni dell' Altana), pp. 89–93.

(1997a) 'It's Not Easy to Exorcise Ghosts', *Common Knowledge* (6), 2, pp. 98–103.

(In press) *Briefwechsel. Paul K. Feyerabend und Hans Albert.* Wilhelm Baum (ed.) (Frankfurt am Main: Fischer).

TRANSLATIONS

(1957) *Die offene Gesellschaft und ihre Feinde, Band I. Der Zauber Platons* by Karl R. Popper. Translated by P. K. Feyerabend (München and Bern: Francke Verlag). 7th edition 1992, reworked (Tübingen: J. C. B. Mohr (Paul Siebeck)).

(1958) *Die offene Gesellschaft und ihre Feinde, Band II. Falsche Propheten: Hegel, Marx und die Folgen.* Translated by P. K. Feyerabend (München and Bern: Francke Verlag). 7th edition 1992, reworked (Tübingen: J. C. B. Mohr (Paul Siebeck)).

(1974) *Kritik und Erkenntnisfortschritt. Abhandlungen des Internationalen Kolloquiums über die Philosophie der Wissenschaft, London 1965, Bd. 4.* Imre Lakatos and Alan Musgrave (eds.). Translated by P. K. Feyerabend and A. Szabó.

Name index

Agassi, J. 152n, 158n
Althusser, L. 168n
Anaxagoras, 74
Anaximander, 52, 55, 73, 74
Aristotle 2, 52, 54, 60, 74, 89, 117–20, 136,
    143, 144, 146 and n, 152n, 153n, 154n,
    164, 172n, 173 and n, 190, 192, 202,
    213
Austin, J. L. 53, 204n
Ayer, A. J. 34

Bacon, F. 52, 104–5, 120, 128n, 186, 202
Barrow, I. 150
Berkeley, G. 25, 150n
Blackmore, J. 13n
Bohm, D. 81n, 84, 215
Bohr, N. 10, 147, 189, 217
Boltzmann, L. 10, 216
Born, M. 84 and n, 122, 148n, 151 and n,
    162
Brahe, T. 203n
Brecht, B. 10, 77, 138, 192–4, 196, 199
Bridgman, P. W. 157n
Bruno, G. 144n, 155n
Bultmann, R. 143 and n
Butterfield, H. 60n

Carnap, R. 41 and n, 88, 109, 133, 136 and n
Cartwright, N. 12n
Clagett, M. 60n, 154n
Cook, J. M. 60n
Comte, A. 166n
Copernicus, N. 59, 113, 119, 121, 144n, 152,
    153n, 154 and n, 155n, 156 and n, 173,
    174, 176, 177, 186, 190, 193

Darwin, C. 216, 219
Darwinists, 184
de Broglie, L. 84
Democritus 175
de Santillana, G. 143n, 144n, 206n
Descartes, R. 52, 53, 98, 104–5, 128n, 202,
    213
Dingler, H. 130n, 132 and n
Diogenes Laërtius, 145

Dirac, P. A. M. 136, 145
Dubislav, W. 110
Duhem, P. 10, 12, 85 and n, 89n, 152 and n,
    155n, 162, 216

Edgley, R. 82
Edison, T. A. 126
Ehrenfest, P. 140n, 147n, 157n
Ehrenfest, T. 140n
Ehrenhaft, F. 92 and n
Einstein, A. 10, 52, 55, 60, 79, 88 and n, 93
    and n, 123, 141, 147 and n, 148n, 157n,
    162, 189, 200, 217 and n
Empedocles, 74
Engels, F. 181
Evans-Pritchard, E. E. 66n, 143n
Exner, F. M. 142n

Faraday, M. 79, 128n
Feigl, H. 39, 81n, 148n, 170 and n
Forman, P. 124n
Freud, S. 66
Fuller, S. 13n
Fürth, R. 92–3 and n

Galileo 2, 10, 52, 59, 60, 79 and n, 85, 89,
    105, 110, 113–23, 127, 128n, 136,
    144n, 149n, 153–6 and n, 169, 170,
    172n, 176, 192–4, 196, 213
Goethe, J. W. von 128n,. 140
Goldberg, S. I. 140n, 147n
Groenewold, H. J. 80–1n

Hanson, N. R. 91n, 137
Heckmann, O. 139, 140n, 152
Hegel, G. W. F. 11, 37, 89n, 131, 162 and n,
    183, 200, 202, 204, 215–16
Heidegger, M. 10n
Heilbron, J. L. 124n, 147n
Heisenberg, W. 54–5, 73, 84, 91n
Helmholtz, H. von 10, 150n, 174, 216
Hempel, C. G. 82 and n, 129n
Heraclitus, 74, 175
Herodotus, 54
Hertz, H. 216

Hesiod, 207n
Hesse, M. B. 144n
Hill, E. L. 82n, 88n
Holton, G. 14, 147n, 148n, 218
Homer, 67, 200
Hume, D. 128, 184
Huxley, A. 58 and n

Ibsen, H. 181
Ionesco, E. 10

Jammer, M. 147n
Jaspers, K. 58n, 98n, 143n
Jung, C. 144n

Kant, I. 2, 53, 109, 128, 130, 202, 203, 204n,
    223n
Katz, D. 24 and n
Kaufmann, W. 147n
Kepler, J. 59, 85, 114, 115, 122, 144n, 150
    and n, 153n, 209
Keynes, J. M. 144n
Kierkegaard, S. 211n
Körner, S. 89n
Koyré, A. 154n
Kraft, V. 42, 87n, 158n
Kropotkin, P. 181
Kuhn, T. S. 8, 11, 14, 61, 70, 81n, 91n, 105,
    124n, 137, 139, 147n, 185, 215, 218

Lakatos, I. 8–9, 11, 14, 109–10, 137, 146n,
    147n, 160–3, 185, 202, 205n, 218–20
Laudan, L. 133 and n, 159n
Lea, H. C. 58n, 96n
Lecky, E. H. 64n
Leibniz, G. W. von 52
Lenin, V. I. 168n, 179 and n
Leucippus, 74
Lévi-Strauss, C. 143n, 144n, 145n, 181
Lysenko, T. 187

McMullin, E. 158n
Mach, E. 10, 12–14, 83–4, 86, 89n, 100,
    129–34, 140n, 196, 216, 217n
Malick, T. 10n
Mao Tse-Tung, 168n
Marshack, A. 206n
Marx, K. 168n, 179, 181, 190
Masters, W. H. 115 and n, 116
Maxwell, G. 30
Maxwell, J. C. 17, 134, 150, 209, 216
Mayer, E. 132–3n
Merton, R. K. 118, 205n
Meyerson, É. 12
Michelet, J. 58 and n
Mill, J. S. 11, 160 and n, 184, 212–16, 224
Miller, D. C. 148 and n

Montaigne, M. de 224
Naess, A. 82
Nagel, E. 83 and n, 86–7, 123, 129n
Neurath, O. 133
Newton, I. 17, 59, 60, 85, 89, 91n, 97–8, 99
    and n, 105, 110, 114–16, 122, 123, 124
    and n, 128, 129n, 132, 133, 136, 137n,
    139–45, 146 and n, 147n, 149–50,
    161–2, 184, 200, 202, 205 and n, 209,
    212, 215, 219

Osiander, N. 154–5n

Palter, R. 152n
Paracelsus 126
Parmenides, 74, 76, 157n
Peirce, C. S. 110
Perrin, J. 93 and n, 141
Planck, M. 10, 89n, 134
Plato, 37, 51, 52, 53, 59, 68, 75, 104, 160,
    176, 195–6
Plutarch, 145
Poincaré, H. 12, 147n
Polanyi, M. 14, 209n, 211n, 218–20
Popper, K. R. 2, 7n, 8, 11, 12, 13, 14, 42,
    52n, 71n, 79n, 81n, 82 and n, 85n, 93n,
    110, 155n, 158n, 159n, 160n, 161, 162,
    164n, 184–5, 202
Post, H. 133n
Pre-Socratics, the 6, 10, 50–77, 188, 200
Ptolemy, C. 144, 154n, 156 and n, 190, 213
Putnam, H. 146n
Pythagoras, Pythagoreans 114, 155, 186,
    190

Quine, W. v. O. 87

Reichenbach, H. 146–7n
Ronchi, V. 128n, 137, 150n, 173n
Rosen, E. 144n, 153n, 154n, 155n
Rosenfeld, L. 84n, 94, 100n, 151 and n
Rubin, E. 33n
Russell, B. A. W. 40–1, 47, 202
Rynin, D. 82, 87n

Sabra, A. I. 128n
Sahlins, M. 206n
Sargant, W. 61, 64n
Scheffler, I. 135n
Schlick, M. 37, 38
Schrödinger, E. 129, 157n
Schücking, E. 139, 140n, 152
Scriven, M. 87n
Sophists, 146, 188
Stokes, G. 14n
Summers, M. 96n
Svedberg, T. 93

ter Haar, D. 84n, 140n
Thales, 52, 53–6, 60, 61, 67–8, 73, 74
Tranekjær-Rasmussen, E. 33 and n, 34
Trevor-Roper, H. R. 143n, 177n
Trotsky, L. 168n

Velikovsky, I. 118
Vienna Circle, the 11, 12, 133 and n, 184
Vigier, J-P. 84
von Dechend, H. 144n, 206n
von Neumann, J. 145, 156–7
von Smoluchowski, M. 93, 140n, 148n

Wagner, R. 72
Watkins, J. W. N. 82
West, M. 99n
Westfall, R. S. 128
Whewell, W. 146n
Whorf, B. L. 44 and n
Wigner, E. P. 157n
Wittgenstein, L. 8, 11, 38, 75, 160n, 208n, 209n

Yates, F. 144n

Subject index

absurdity 75, 101–2, 106, 145, 175
acupuncture 186, 220
alchemy 205
anarchism 123, 176
'anything goes' 123, 218
arguments(s) 10
art(s) 7 and n, 10, 105–6, 192–9, 206n
astrology 120, 125, 161, 190, 219
atomic theory 41–2, 73, 74, 132, 133, 138,
    142, 147, 172, 205–6, 216
autonomy principle 91–4

Brownian motion 92–3, 141–2, 148n

children, childhood 104, 105, 110, 119,
    188–9, 208, 211
Church Fathers, the 63, 75, 104, 200
closed societies, 62, 68, 69
condition of meaning invariance 83–9,
    97–9, 101
confirmation 79, 91, 95, 99, 101, 135, 136,
    170–1, 197, 218
consistency condition 83–97
contexts of discovery and justification, 171,
    176
criticism(s) 51, 55–6, 64, 66, 68, 69–72, 76,
    77, 80, 81, 100, 102–3, 111, 128, 129,
    130, 131, 132, 133, 134, 143 and n, 144,
    145, 163, 164, 170, 182

decision(s) 24, 31, 33, 38, 42, 62, 71, 77,
    158n, 167, 179, 208
democracy 9, 13, 78–9, 187, 212–26
democratic relativism 9, 220–6
demythologization 6, 143 and n, 182
dialectical materialism 202
dilettantes, 112, 118, 122
dogma, dogmatism 43, 53, 62, 63, 66, 68, 70,
    71, 72, 74, 77, 78–82, 94, 99, 102, 127,
    133, 160, 183, 187, 196
doubt 25–35, 41, 106, 170

education 97, 119, 182, 188–91, 195, 207,
    214, 215, 218, 219, 223
élitism 14, 119, 219 and n, 220, 221, 222

empirical content 3, 91–5, 102, 130, 135,
    136, 141, 142, 143n, 162, 175, 198, 200
empiricism 1–2, 4, 56, 58, 61, 71, 76,
    78–103, 119–20, 122, 128, 132, 133,
    136, 148n, 150, 152n, 168, 174, 175,
    197, 214, 215, 224
enlightenment 96, 177n, 181, 182, 203,
    223n, 224
ethics 5, 14, 50, 62–3, 66–72, 78, 179
evidence, 2, 3, 30, 33, 36–7, 42, 56, 65, 70,
    94, 95, 106, 119, 121, 135, 137, 144n,
    151, 163, 165, 166, 173, 175, 206, 218
evolution, evolutionary theory 6, 106, 111,
    119, 143, 187, 220
experience 18, 36–7, 40, 46–9, 56, 57, 58,
    65, 76, 79, 80, 88, 94, 98, 100, 102, 109,
    117, 119, 120, 122, 124, 128, 132, 136,
    139, 145, 146, 149, 150, 154, 155, 158,
    163, 169, 173, 175
  unity of 49
experiment(s) 1–2, 19, 27, 32, 33, 38, 39, 41,
    47, 56, 59, 78, 80, 81, 86, 105, 108, 109,
    121, 124, 132, 141, 142, 146, 147 and n,
    153n, 164, 165, 169, 175, 207, 212
experts 106–7, 112–26, 186, 187, 188, 189,
    205, 218, 221, 223
explanation 17, 21, 31, 55, 56, 64–8, 76,
    82–7, 94, 95, 96, 105, 121, 158, 161,
    169, 175, 206, 212

falsification(s), falsificationism, falsifiability
    110, 137, 146 and n, 158–9, 163–4,
    184–5, 203
film, *see* movies
free society 112, 125, 127, 181–91, 218, 220,
    222

gravitation 17, 44, 60, 85, 146 and n, 212,
    215
'grue' 127, 135, 197

humanitarianism 12, 63–4, 67, 69, 97, 111,
    126, 191

impetus theory 97–8, 121

indoctrination 61–2, 64, 182
induction 56, 60, 99, 100–1, 139, 145–6,
    156, 159, 214, 216, 218
instruments 21, 22, 28, 38, 108, 113, 117,
    123, 141, 168, 169, 172–4, 182, 194,
    210, 214, 223, 225

justification 21, 22, 100, 145, 184

knowledge 3, 4, 5, 6, 7, 12, 17, 28, 38, 44,
    50, 51, 52, 53, 55, 68, 69, 70, 71, 72, 74,
    76, 77, 78, 80, 83, 89, 98, 99, 100, 101,
    102, 104, 105, 106, 107, 109, 111, 118,
    122, 124, 125, 126, 127, 129, 130, 133,
    134, 145, 158 and n, 160, 164, 166, 167,
    168, 170, 171–7, 178, 184, 194, 195,
    197, 200, 202–3, 212–18, 220, 224

language(s), artifical or 'ideal' 23, 38, 134–5,
    217
  'ordinary' 19, 23, 38, 42–6, 75, 79n, 101,
    176
logical reconstruction 11, 134–5

magic 119, 120, 125, 143n, 190, 205
Marxism 4, 66, 164n, 172, 176, 182, 198,
    214
meaning 39–40, 47, 49, 81, 82–8, 96, 97–9,
    101, 102, 184, 213
medicine 125–6, 164n, 178n, 179, 186–7,
    195, 207, 215, 220–5
metaphysics 2, 3, 78–82, 95, 99–100, 102,
    109–10, 132, 143, 152, 175, 179, 197
method, methodology 6, 7, 8, 9, 17, 18, 21,
    42, 48, 49, 55, 80, 89, 92, 96, 97, 100,
    107, 122, 123, 124, 125, 128, 129–37,
    138–80, 183–7, 197, 206, 212–17
monism, theoretical 2, 4–6, 14, 99, 104–7
movies 199
myth 1–4, 8, 55–72, 76, 77, 95–7, 105, 124,
    143–4 and n, 163, 165, 188, 195,
    198–9, 207n
myth predicament, the 2–4, 95–7, 214

'normal science' 8, 185, 215

observation 16–49
opportunism 121, 122, 123, 217 and n
optics 45, 85, 110, 121, 127, 128, 153, 169,
    174, 186, 196

pain(s) 23–8, 33–7
parapsychology 201, 219
philosophy of science 7–8, 77, 95, 127–37,
    140n, 146n, 156, 158, 172, 176, 185,
    196–9, 217, 218
physiology 20, 22, 31, 101, 110, 115, 116,

    131, 138, 145, 158, 167, 168, 170, 172,
    174, 225
pluralism, methodological 216–17
pluralism, theoretical 4–6, 7, 75, 80–1,
    92–9, 102–3, 107–11, 145, 213
progress 6, 60, 62, 69, 73, 75, 76, 78, 79, 83,
    87, 90, 91, 98, 99, 101, 110, 122, 124,
    125, 135, 139, 142, 143, 161, 168, 170,
    175, 176, 177, 185, 187, 189, 194, 198,
    218
proliferation 105, 106, 107–11, 215, 217,
    224
psychoanalysis 4, 64–5, 66, 172, 219

quantum mechanics 4, 66, 76, 80, 82n,
    84–5, 91n, 93–5, 102n, 125, 128n, 129,
    137, 140, 142 and n, 144n, 151 and n,
    155n, 157 and n, 163, 172, 214, 215

rationalism, rationality 14, 60, 61, 68, 122,
    161, 176, 177, 188, 192, 194, 200–11,
    212, 214, 215, 216, 223, 224
realism 19, 46, 48, 130
  naïve 16, 169–73
reason 10, 14, 62, 150, 156, 158, 164, 175,
    176, 177 and n, 189, 192, 194, 195, 196,
    204, 209, 210, 215, 216, 218
relations, relationships 44, 45, 48, 86, 87
relativism 5, 9–10, 13, 14, 208, 220–6
relativity, theory of 40, 60, 85, 86–7, 88, 125,
    129, 134, 135, 137, 147 and n, 148 and
    n, 163, 183, 184, 209, 219
religion 1, 60, 63, 71, 72, 75, 119, 125, 130,
    134, 149, 181, 182, 187, 188, 190, 202,
    220
results of science 8, 128, 183, 186–8, 194,
    201, 207, 215, 221

'schizophrenia', methodological 124, 128,
    137 and n
scientific revolution, the 10–11, 119, 127,
    194, 196, 206, 213
sensations 24–38, 45, 48, 49, 56, 100, 130–4
sense-data 22–39, 40, 47, 71, 109, 132, 133
simplicity 91, 135
stability thesis 39–44
standards 14, 69, 113, 127, 166, 179, 184,
    195, 201–3, 205, 207–11, 212, 216–22
'subjective visual grey' 24

tenacity, principle of 6, 107–8, 111, 143,
    174–5, 215
test(s) 18, 40, 43, 92, 93, 102, 110, 136, 157,
    159n, 210, 212
theatre 9, 192–9
theoretical entities 16–49
thermodynamics 83, 85, 108, 140

'thing-language' (Carnap) 41, 42, 46, 88,
    109, 133
tradition(s) 43, 57, 74, 122, 160, 167, 189,
    203–4, 207–10, 213, 220–6
trial by jury 106, 187
truth 3, 4, 5, 6, 8, 40, 42, 46–9, 57, 62–8,
    70, 72, 75, 76, 95, 96, 97, 104, 107, 118,
    121, 126, 145, 160, 170, 179, 182–3,
    189, 194, 198, 200, 213, 214, 224

values 14, 174, 223
verification, verifiability 18, 134, 144n, 147n,
    158–9, 160, 163
voodoo 145 and n

witchcraft 2, 4, 58n, 60, 95, 98,, 120, 125,
    126, 136, 143–4n, 166, 177 and n, 178,
    186